IAS/PCMI—The Teacher Program Series
Mathematics for Teaching: A Problem-Based Approach

Volume 4

Some Applications of Geometric Thinking

Bowen Kerins • Darryl Yong
Al Cuoco • Glenn Stevens • Mary Pilgrim

American Mathematical Society
Institute for Advanced Study
Park City Mathematics Institute

Mathematics for Teaching: A Problem-Based Resource for Teachers was developed at Education Development Center, Inc. (EDC) in partnership with the Park City Mathematics Institute (PCMI), with the support of the National Science Foundation.

Development Team: Bowen Kerins, Darryl Yong, Al Cuoco, Glenn Stevens, Mary Pilgrim

2010 *Mathematics Subject Classification.* Primary 00-01; Secondary 00A07.

For additional information and updates on this book, visit
www.ams.org/bookpages/sstp-4

Library of Congress Cataloging-in-Publication Data

Mathematics for teaching : a problem-based approach / Bowen Kerins [and three others]. volumes cm.—(IAS/PCMI, the teacher program series)

"Institute for Advanced Study."

"Park City Mathematics Institute."

"Developed at Education Development Center, Inc. (EDC) in partnership with the Park City Mathematics Institute (PCMI), with support of the National Science Foundation."

Based on a summer series of workshops held in 2009.

Includes bibliographical references.

Contents: volume 1. Probability through algebra—volume 2. Applications of algebra and geometry to the work of teaching—volume 3. Famous functions in number theory—volume 4. Some applications of geometric thinking

ISBN 978-1-4704-2195-3 (alk. paper)

1. Mathematics teachers—Training of—Congresses. 2. Mathematics—Study and teaching—Congresses 3. Algebra—Congresses. 4. Probabilities—Congresses. 5. Number theory—Congresses. I. Kerins, Bowen, 1975– II. Institute for Advanced Study (Princeton, N.J.) III. Park City Mathematics Institute.

QA11.A1M23445 2015
510.71–dc23 2015022685

Others who contributed include: Gail Burrill, James King, Alison Langsdorf, Ryota Matsuura, and Kevin Waterman.

∞ The paper used in this book is acid-free and falls within the guidelines
established to ensure permanence and durability.
Visit the AMS home page at `http://www.ams.org/`

10 9 8 7 6 5 4 3 2 1 21 20 19 18 17 16

Contents

Preface

"To Think Deeply about Simple Things" was a motto of Arnold Ross, founder of the Ross Mathematics Program. Ross's philosophy has provided inspiration for many content-based professional development programs, each with its own personality and design, including the Summer School Teacher Program (SSTP) at the Park City Mathematics Institute (PCMI) and the PROMYS for Teachers Program (PfT) at Boston University.

More information about the SSTP can be found at `mathforum.org/pcmi/sstp`. For PfT, see `www.promys.org/pft/`.

PCMI, sponsored by the Institute for Advanced Study, is a three-week summer program for those involved in mathematics: research mathematicians, graduate students, undergraduate faculty, undergraduate students, and precollege teachers.

The SSTP has been an integral part of PCMI from the beginning. Since 2001, the mathematics course has been designed for precollege teachers by a collaborative of teachers, educators, and mathematicians from PfT. The SSTP course meets for 15 two-hour sessions during which the teachers investigate an aspect of mathematics loosely related to the overall mathematical theme of that summer's PCMI.

This book is based on the program from 2011, *Some Applications of Geometric Thinking*. One goal of this course is to help teachers see that geometric ideas can be used throughout the secondary school curriculum, both as a hub that connects ideas from all parts of secondary school and beyond—algebra, number theory, arithmetic, and data analysis—and as a locus for applications of results and methods from these fields. The connection to PCMI's overall mathematical theme of 2011 (Moduli Spaces of Riemann Surfaces) is through a focus on parametrization and geometric representation.

Another goal, one that runs across all the summers, is to take up mathematics that helps teachers put the topics in their curricula into the broader landscape of mathematics as a scientific discipline.

The Design of the Course

The central feature of the SSTP course is a set of intricately sequenced questions that engage participants in doing mathematics in ways that exemplify the *Common Core* Standards for Mathematical Practice. What's important here is the structure of the program, because the program is not a "course" in the traditional sense. The materials provide participants with the opportunity for authentic mathematical discovery—participants build mathematical structures by investigating patterns, use reasoning to test and formalize their ideas, offer and negotiate mathematical definitions, and apply their theories and mathematical machinery to solve problems. Through this experience, participants develop habits of mind for thinking about and doing mathematics, deepening their mathematical intuition, sense-making, and reasoning skills.

Each fall and spring, a team drawn from the PfT community meets regularly to create two or three major themes for the upcoming SSTP course, and for each theme, it creates a "soup" of potential problems and investigations that might be used at PCMI. Once in Park City, the instructors create daily problem sets, revised each night to reflect what happened in the day's session. After the course, the problem sets are revised once more, solutions and hints are written, and the course is prepared for publication.

The problem sets are separated into three sections: Important Stuff, Neat Stuff, and Tough Stuff. The problems in Important Stuff contain the fundamental concepts that should enable everyone to move forward. The problems in Neat Stuff and Tough Stuff are there for those who are curious or looking for a challenge.

The distinguishing features of the program have stayed constant over the years:

Teachers as professionals These materials are designed and implemented by practicing teachers in collaboration with mathematicians and mathematics educators. Experienced teachers mentor teachers new to the program by acting as "table leaders." The connections between the program and the teaching profession are real, because teachers are involved at every level.

In one summer at the SSTP, teachers studied how formal algebra with polynomials can be used to bring coherence to combinatorial problems. In another summer, teachers learned to apply the arithmetic of algebraic numbers to the problem of designing tasks for their students that "come out nice."

Serious mathematics connected to secondary teaching Each experience is designed to connect to the mathematics teachers use in their professional lives. This "applied" mathematics is sometimes around underpinnings that will end up in the hands of students. But it may also take up mathematics that helps teachers put the topics in their curricula into the broader mathematical landscape.

Experience before formality Participants experience first-hand the effectiveness of struggling with new ideas and connections *before* they are brought to closure. The role of the instructor is to pull together the participants at several points to collate conjectures,

> logical arguments, and extensions. With help from those around them, teachers refine and prove their own conjectures, sometimes over the course of several days. This style of learning, emblematic of the intent of the *Common Core's* Standards for Mathematical Practice, has had an immense effect on how teachers approach their own classes and how they view the discipline they teach.

Teachers work in tables of 5–6 participants, accompanied by a table leader who typically responds to questions with well-posed additional questions. Tablemates work on carefully crafted problem sets, slowly abstracting general principles from calculations and experiments.

The goal of the PCMI teacher program is to provide teachers with opportunities to:

- deepen their understanding of mathematics,
- reflect on the practice of teaching, and
- serve as mathematical resources for their colleagues.

All of this has worked. Exit reviews of the summer programs and in-depth interviews by external evaluators of teachers from varying backgrounds and school systems have provided evidence that the program has helped hundreds of teachers become more effective in—and more satisfied with—their professional lives. According to one evaluation report, many participants have been influenced by PCMI to revise their roles as teachers, acting more as facilitators, rebalancing how much time they allow students to talk versus talking themselves, and giving students more responsibility for their own learning. They consciously change the ways they question students and answer questions, and make reasoning and sense making a core goal of their instruction.

Teachers meet after each mathematical session to discuss the work of teaching. One summer focused on how to make classrooms a place where questioning is central to learning. In another summer, participants considered how to manage discourse that made students the center of the discussion. Activities are designed around artifacts of practice such as student work, classroom videos, assessment, or lesson design.

While these materials were developed for use at the SSTP, they have been piloted in other settings and have been successfully adapted for use as a capstone course for preservice mathematics teachers and as an elective in a mathematics or mathematics education program where students did one problem set per week. While the course does not replace standard advanced geometry or complex analysis courses, it prepares students to take such courses, with the problem sets building towards a mathematical conclusion. The materials could be used as a course in a graduate program or summer institute for teachers, or in a variety of different professional development configurations. Intermediate conclusions typically occur every four or five problem sets, so a shortened course might consist of the first four or five problem sets or the eighth or ninth problem sets. In settings where credit is offered, some instructors have assigned projects or reports on specified readings related to the work. While the mathematics may seem appropriate only for teachers with a strong mathematical background, it was designed to provide all teachers with a rich mathematical experience, regardless of their backgrounds.

Navigating the Problem Sets

To allow access for multiple levels of experience, any topic in algebra or beyond is approached as though it is newly encountered. For example, the work on quadratics is designed to lead to an understanding of the quadratic formula, not requiring it, and generalizations come from repeated iterations of examples, not from an initial exposure. Many learners find their first answers by testing options, and when an unusual one comes along, they use the patterns from their previous work to figure out what might happen in this new case. In some cases, the proof uses all of the key ideas from the openers (boxed problems). The problems are not built to support a lecture but rather deliberately constructed for students to pursue a general solution over time, with the goal of enabling the learners to build their own understanding from the problems as they work through them.

Note that because the problems are carefully crafted to let the mathematics unfold through the experience of actually doing the work, facilitators should do all of the problems, at least in Important Stuff, themselves before engaging students so as to anticipate students' thought processes and to encourage them during explorations and discoveries.

A few things are important about how the materials are structured:

- The boxed openers are meant to be answered just like the rest of the problems. Consider them "Problem 0"; they are boxed because they are intended to be more important than others. A general note: keep looking forward in the problem sets. The solutions do not go immediately for proof, but rather the course tends to let ideas sit for a while. For example, a proof of the Set 3 opener might be found in a problem in Set 4.
- There are problem categories: *Important Stuff*, *Neat Stuff*, *Tough Stuff*, and maybe other stuff sometimes. All the mathematics that is central to the program

can be found and developed in the Important Stuff. That's why it's Important Stuff.

- As mentioned earlier, the materials provide experience before formality, where the participant uses examples to build intuition. Definitions and theorems appear as capstones, not foundations.
- The problems should lead to the appropriate mathematics rather than requiring it. The same goes for technology: the problems should lead to the appropriate use of technology rather than requiring it.
- The problems have multiple points of entry where everyone, regardless of the level of confidence or experience, can begin.
- The materials have a low threshold and a high ceiling—they allow everyone to experience success and are also designed so that participants will feel challenged regardless of skill or experience level.
- The problems explicitly link different content areas and encourage participants to seek multiple representations and solutions.
- A problem presented from one perspective (algebraically, say) may be repeated in another (geometrically, for example).
- The problems often foreshadow key ideas that are not introduced formally until later problem sets.
- The problems repeat and connect throughout all of the sets. A goal is to have the participants look for connections rather than being surprised when they notice relationships.

CHAPTER

1 Problem Sets

Welcome to the course.

We know you'll learn a great deal of mathematics here—maybe some new tricks, maybe some new perspectives on things with which you're already familiar. A few things you should know about how the class is organized:

- **Don't worry about answering all the questions.** If you're answering every question, we haven't written the problem sets correctly.

Some of the problems have yet to be solved. It's especially good to think about these.

- **Don't worry about getting to a certain problem number.** Some participants have been known to spend all of their effort working on just one problem from a problem set (and perhaps a few of its extensions or consequences).
- **Stop and smell the roses.** Getting the correct answer to a question is not a be-all and end-all in this course. How does the question relate to others you've encountered? How do others think about this question?
- **Respect everyone's views.** Remember that you have something to learn from everyone else. Remember that everyone works at a different pace.
- **Learn from others.** Give everyone a chance to discover, and look to those around you for new perspectives. Resist the urge to tell others answers if they aren't ready to hear them yet. Try to avoid using technology to solve a problem "by itself." There is probably another, more interesting, way.
- **Each problem set has its Stuff.** There are problem categories: Important Stuff, Neat Stuff, Tough Stuff, and maybe other stuff sometimes. Check out Impor-

> tant Stuff first. All the mathematics that is central to the course can be found and developed in the Important Stuff. *That's* why it's Important Stuff. Everything else is just neat or tough.

When you get to Problem Set 3, come back and read this introduction again.

Problem Set 1

Opener

Get geometry software working on your computer. The actual opener will be done later.

There are several geometry software packages you can use, but if you are working with others, it is best to standardize on one. Some provided files have been built for Geometer's Sketchpad.

Important Stuff

1. Sameer tells you the perimeter and area of a rectangle. Is it possible to confidently determine the dimensions of the rectangle?

2. Here are some perimeters and areas. Find the dimensions of each rectangle.
   - **a.** perimeter 24, area 36
   - **b.** perimeter 24, area 35
   - **c.** perimeter 24, area 32
   - **d.** perimeter 24, area 27
   - **e.** perimeter 24, area 34
   - **f.** perimeter 24, area 37

3. Find all solutions to each equation.
   - **a.** $x^2 - 12x + 36 = 0$
   - **b.** $x^2 - 12x + 35 = 0$
   - **c.** $x^2 - 12x + 32 = 0$
   - **d.** $x^2 - 12x + 27 = 0$
   - **e.** $x^2 - 12x + 34 = 0$
   - **f.** $x^2 - 12x + 37 = 0$

4. A rectangle has perimeter 36. What could its area be?

5. Chance tells you the surface area and volume of a rectangular box. Is it possible to confidently determine the dimensions of the box?

Opener

Three cities located at points M, A, L get together to build an airport. Where should the airport be placed to minimize the lengths of the new roads that need to be built?

In geometric language, the goal is to find point X so that $MX + AX + LX$ is minimized.

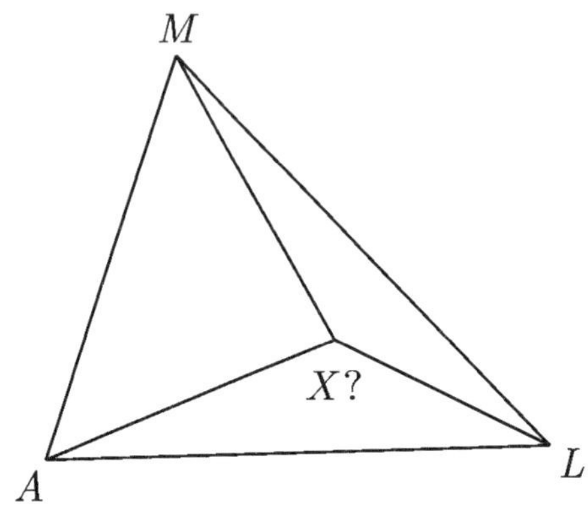

Build this sketch, then use geometry software to figure out where point X should be placed. Is there anything special about this point?

Neat Stuff

6. For each point, decide if it is the same distance from $(7,1)$ and $(-2,9)$.

a. $(7,10)$ **b.** $(-1,1)$ **c.** $(-17,-17)$ **d.** $(-2,0)$

This problem has four parts, but it's *not* multiple choice!

7. Find some rectangles whose perimeter and area have the same numeric value. Then find some more.

8. Find some rectangular boxes whose surface area and volume have the same numeric value. Then find some more, if there are any.

9. Rina tells you the perimeter and area of a triangle. Is it possible to confidently determine the side lengths of the triangle?

10. On July 4, 2011, the date was written as 7/4/11, and $7 + 4 = 11$.

a. After July 4, 2011, how many more times in the 21st Century is there a day like this?

b. How many times *next* century will there be a day like this?

c. How can your second answer help you check the first?

The next one after 7/4/11 was 8/3/11.

Tough Stuff

11. A triangle has perimeter 24. Find its maximum possible area, and explain how you know that this *must* be it.

12. Given positive integer n, the unit fraction $\frac{1}{n}$ can be written as the sum of two other unit fractions:

$$\frac{1}{n} = \frac{1}{a} + \frac{1}{b}$$

Like the blood type, a and b must be positive. Unlike the blood type, they must be integers.

Find a rule for the number of ways to write $\frac{1}{n}$ as the sum of two unit fractions.

13. Find a rule for the number of ways to write $\frac{1}{n}$ as the sum of *three* unit fractions.

Problem Set 2

Opener

Art (at point A) has to take a high-stakes test at point T. He is extremely thirsty, and needs to take a big gulp from nearby river NL.

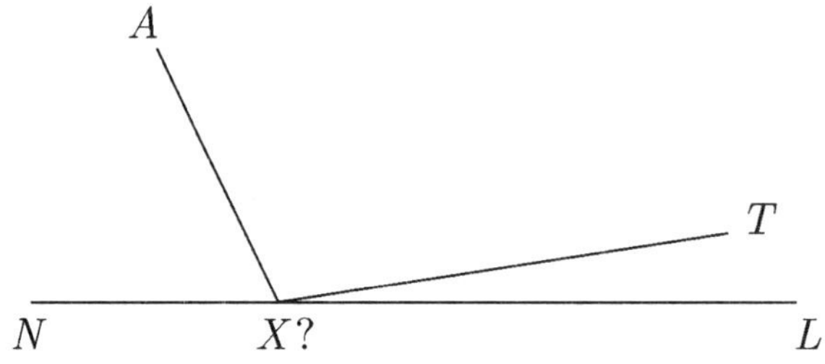

If you figure out where point X should be, start thinking about how you could *construct* point X more directly, how you could *prove* that point X must be the right one, or how to change the problem in interesting ways.

Where should he run toward (point X) to minimize the total distance $AX + XT$?

Important Stuff

1. **a.** What's $(11+\sqrt{2})+(11-\sqrt{2})$?
   b. What's $(11+\sqrt{2})(11-\sqrt{2})$?
   c. What's $(11+x)(11-x)$?

2. **a.** A rectangle has perimeter 44 and area 117. What are its length and width?
   b. A rectangle has perimeter 44 and area 119. What are its length and width?
   c. A rectangle has perimeter 44 and area 100. What are its length and width?
   d. A rectangle has perimeter 44 and area 130. What are its length and width?

3. Find all solutions to each quadratic equation.
   a. $w^2 - 22w + 121 = 0$ **b.** $w^2 - 22w + 120 = 0$
   c. $w^2 - 22w + 117 = 0$ **d.** $w^2 - 22w + 119 = 0$
   e. $w^2 - 22w + 100 = 0$ **f.** $w^2 - 22w + 2 = 0$
   g. $w^2 - 22w - 408 = 0$ **h.** $w^2 - 22w + 130 = 0$

We like shortcuts, but we *especially* like shortcuts that do not involve the phrase "$4ac$".

4. Two numbers add up to 200 and their product is 9,991. What are the numbers?

5. Find three rectangles whose perimeter, in units, is the same number as the rectangle's area, in square units.

6. Triangle SAM has points $S(2,1)$, $A(4,1)$, and $M(4,6)$.
   a. Draw triangle SAM in the plane.
   b. New points are created from the points in triangle SAM according to the rule

   $$(x,y) \mapsto (-y,x)$$

   Draw the new triangle created in this way in the same plane, and describe how this triangle is related to the original.

7. Transform triangle SAM according to the rule

$$(x,y) \mapsto (-x,-y)$$

Draw the new triangle and describe how this triangle is related to the original.

8. a. Show, beyond a reasonable doubt, that $(14,-4)$ is *not* equidistant from $(-2,-5)$ and $(4,11)$.
   b. Again for $(-10,7)$.
   c. Build a rule you could use to decide whether (x,y) is equidistant from $(-2,-5)$ and $(4,11)$.

9. Cuong repeats the transformation in Problem 6 a whole bunch of times.
   a. What happens after the transformation is applied twice?
   b. What happens after the transformation is applied three times?
   c. . . . four times?
   d. . . . five times?
   e. . . . thirteen times?
   f. . . . 101 times?

Neat Stuff

10. For any rectangle you can assign a point (l,w) in a coordinate plane, defined by the length and width of the rectangle.
    a. Plot four points that all correspond to rectangles with area 20.
    b. How many rectangles are there with area 20? Plot them all.
    c. Plot all the rectangles with perimeter 20.
    d. Is there a rectangle with perimeter 20 *and* area 20?

11. Repeat Problem 10 for rectangles with area 12 and perimeter 12. What happens?

12. **a.** Put a point Q on the number line. Define $f(P)$ to be the distance from point P on the number line to point Q. What does the graph of $f(P)$ look like?

b. Put point R on the number line, and now define $f(P)$ as the total distance from any point P to the two given points. What does the graph of $f(P)$ look like?

c. Put point S on the number line, and do all that stuff we said again.

13. Find three triangles that have the same numeric value for their perimeter and area.

14. Let $f(x) = x^3 + 3x^2 - 8x - 80$. Use long division to find the remainder when $f(x)$ is divided by each of the following.

a. $(x - 1)$
b. $(x - 2)$
c. $(x - 3)$
d. $(x - 4)$
e. $(x - 5)$

15. Complete this table for $f(x) = x^3 + 3x^2 - 8x - 80$.

x	f(x)
0	
1	
2	
3	
4	
5	

16. Find the remainder when $f(x) = x^{12} + 3x - 1$ is divided by each of the following.

a. $(x - 1)$ **b.** $(x - 2)$ **c.** $(x - 10)$ **d.** $(x + 1)$

17. Jessica takes triangle SAM from Problem 6 and applies a wacky transformation:

$$(x, y) \mapsto (x + y, -3x + 7y)$$

a. Draw this new triangle JES. Is it even a triangle anymore?

b. What is the area of this new shape? How does JES compare (in area) to SAM?

18. Find all the ways to write $\frac{1}{10}$ as the sum of two unit fractions. Here's one for free:

$$\frac{1}{10} = \frac{1}{20} + \frac{1}{20}$$

A *unit fraction* has numerator 1 and positive integer denominator.

Tough Stuff

19. There's a point inside most triangles that forms three 120° angles with segments to the three vertices. A *Matsuura triangle* is a triangle whose side lengths are all integers, *and* whose three interior segment lengths from the 120° point are also integers. Find some Matsuura triangles, or prove they do not exist.

20. Complete this long division problem, where all the missing digits are marked with an X. (There is no remainder.)

```
             X X 8 X X
       ___________________
X X X ) X X X X X X X X
          X X X
        -------
            X X X X
              X X X
            -------
                X X X X
                X X X X
                -------
```

Problem Set 3

Opener

Equilateral triangle DAN has point X inside it.

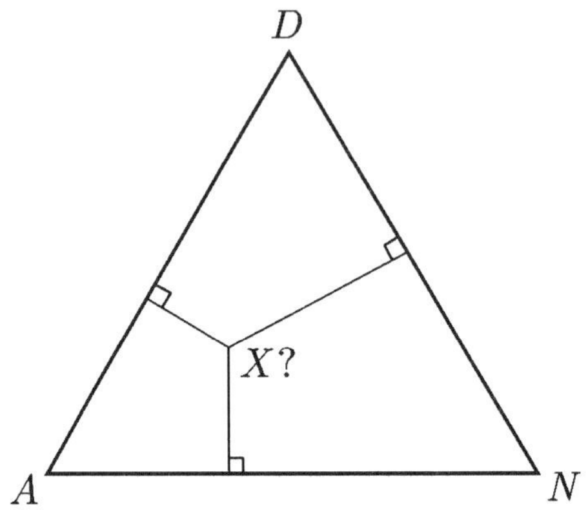

Where should point X be placed so that the three *perpendicular* segments to the three sides of the triangle have a minimum total length?

Important Stuff

1. Two numbers add up to 200, and their product is 9,919. Find the two numbers.

Using geometry software, you can draw a perpendicular line . . . then what?

2. **a.** Two numbers add up to 30, and their product is 161. Find the two numbers.
   b. Two numbers add up to 30, and their product is $225 - n$. Find the two numbers.
   c. Two numbers add up to 30, and their product is 289. Find the two numbers.

Say, isn't 289 a perfect square?

3. For each point, determine whether it is 10 units away from $(13, -7)$.

   a. $(5,3)$ **b.** $(7,1)$ **c.** $(10,-16)$ **d.** (x,y)

4. For any rectangle you can assign a point (l, w) in a coordinate plane, defined by the length and width of the rectangle. Rectangle RAUL has length 10 and width 15.
   a. Plot four points that all correspond to rectangles similar to RAUL.
   b. How many rectangles are similar to RAUL? Plot them all.

Two shapes are *similar* if one is a scaled copy of the other.

  c. Find a rectangle similar to RAUL, but whose perimeter in units is larger than its area in square units.
  d. Find a rectangle similar to RAUL whose perimeter and area have the same numerical value.

5. Darryl is so one-dimensional, he lives on the x-axis. He needs to make a round-trip once each to $(1,0)$, $(3,0)$, and $(17,0)$ from home. Where should Darryl live to minimize his total travel distance? You have 30 seconds to guess . . . go!

29 . . . 28 . . . 27 . . .

6. a. Point Q has coordinates $(3,0)$ and point P has coordinates $(x,0)$. Define $f(P)$ to be the distance from point P to point Q. What does the graph of $f(P)$ look like?
  b. Point R has coordinates $(17,0)$. Now define $f(P)$ as the sum of the distances PQ and PR. What does the graph of $f(P)$ look like?
  c. Point S has coordinates $(1,0)$. You take it from here.
  d. Where should Darryl live to minimize his total travel distance?

7. Find three rectangular boxes where the numeric value of their surface area equals the numeric value of their volume.

One of them is a common shape of Jell-O.

8. Bill bought two briefcases this weekend. Remarkably, both have the same surface area and both have the same volume. And yet, the two briefcases are not the same size. Find some possible dimensions for Bill's briefcases.

One of the briefcases may have been 80% off, but that doesn't really impact the problem.

9. Find two triangles where the numeric value of their perimeter equals the numeric value of their area.

Or not. There may not even be two such triangles.

Neat Stuff

10. Find three pairs of positive numbers a and b, with $a \geqslant b$, that satisfy

$$\frac{1}{a} + \frac{1}{b} = \frac{1}{2}$$

11. Find all rectangles with integer side lengths whose area, in square units, is exactly double the perimeter in units.

12. Let $p(x) = x^4 - 2x^3 + 3$ and $q(x)$ is the *remainder* when $p(x)$ is divided by $(x-2)(x-5)$ . . . in other words, $x^2 - 7x + 10$.

Before calculating, what kind of thing could $q(x)$ be? Be as specific as you can.

a. Complete this table of values for $p(x)$:

x	p(x)
0	
1	
2	
3	
4	
5	

b. Complete this table of values for $q(x)$:

x	q(x)
0	
1	
2	
3	
4	
5	

Notice anything interesting?

13. Let $f(x) = 3x^2 - 10x + 21$. Use polynomial long division to find a linear function that agrees with $f(x)$ when $x = 3$ and when $x = -5$.

14. Let $f(x) = 3x^2 - 10x + 21$. Ooh, the same function.

a. Use division to find $g(x)$, a linear function that agrees with $f(x)$ when $x = 3$ and when $x = 3$.

This isn't a typo. Use the method from Problem 13.

b. Graph $f(x)$ and $g(x)$ on the same axes. What do you notice?

15. Find an equation of the tangent line to $f(x) = x^3$ at $x = 2$ using the method described above.

16. Redo this problem set's opener, using a generic triangle instead of an equilateral one. What happens?

17. **a.** Point A has coordinates $(0, 0)$ and point P has coordinates (x, y). Define $f(P)$ to be the distance PA. What does the graph of $f(P)$ look like?

b. Point L has coordinates $(8, 0)$. Now define $f(P)$ as the sum of the distances PA and PL. What does the graph of $f(P)$ look like?

c. Point M has coordinates $(3, 6)$. You take it from here.

18. In the Opener problem from Set 2, Art runs just as fast to the river as he does away from it. But in reality, Art runs twice as fast when he's thirsty as he does when he's not. How does this change the sketch and its solution? Can you model this with your geometry software? On a graphing calculator? What is Snell's Law?

19. What is the largest prime number ever used in the lyrics of a Top 40 song?

Tough Stuff

20. Find the equation of a parabola that is tangent to the function $f(x) = x^4 - 2x^3 + 3$ at $x = 1$, and also intersects $f(x)$ at $x = -1$, all without using any calculus.

21. Given a positive integer k, there are a number of values of b so that the quadratic $x^2 + bx + kb$ is factorable over the integers. Determine, based on k, how many such values of b there are.

22. The quadratic equation $x^2 - 10x + 22 = 0$ has two roots.

a. Find a quadratic whose roots are the *squares* of the roots of $x^2 - 10x + 22 = 0$.

b. Find a quadratic whose roots are the nth powers of the roots of $x^2 - 10x + 22 = 0$.

Problem Set 4

Important Stuff

1. Use the figure below.

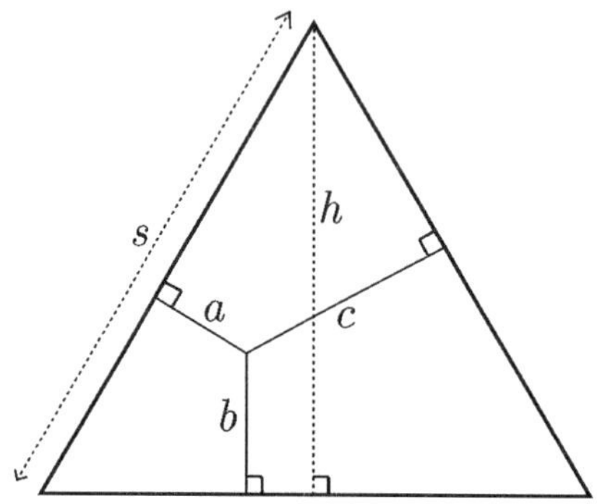

a. Find an expression for the area of the equilateral triangle pictured above.

b. Find a triangle with area $\frac{1}{2}as$. It might be hiding!

c. Find a second expression for the area of the equilateral triangle, then use your expression to prove a result from Problem Set 3.

It's okay for the expression to have more than one variable.

Opener

Use geometry software to sketch a right triangle that stays a right triangle when you move its points around.

What would you need to make first?

Then, build three equilateral triangles on the outside of the right triangle by using each side of the right triangle as the base for an equilateral triangle.

Find a relationship between the areas of the equilateral triangles.

Opener, Part 2

Redo the above construction using an arbitrary triangle (that is, a generic one instead of a right triangle) as the starter, building equilateral triangles along each side.

Use this sketch to *construct* a point inside the triangle that forms three 120° angles when the point is joined to the triangle's three vertices. See if you can figure out how, and why. If you've already built this construction, try to find another. The sketch from Problem Set 3 might help you find a proof!

---

2. List several rectangles whose area and perimeter have the same numeric value. For each rectangle, calculate

$$\frac{1}{L} + \frac{1}{W}$$

where L and W are the length and width of each rectangle.

3. Rectangle ARON has length 20 and width 15.

   a. In a coordinate plane, plot (l, R), where l is the length of ARON and R is the ratio of the area of ARON to the perimeter of ARON.
   b. Find several rectangles similar to ARON, and plot them on the same axes.
   c. Find the one rectangle similar to ARON whose perimeter and area have the same numeric value.

4. Tim starts with a 3 by 4 by 5 box.

   a. Mimi finds a box similar to Tim's box whose surface area and volume have the same numeric value. What are the dimensions of Mimi's box?
   b. Call the dimensions of Mimi's box T, I, and M. Find the exact value of

   $$\frac{1}{T} + \frac{1}{I} + \frac{1}{M}$$

   What!!

5. For each point, determine whether it is twice as far from $(20, 0)$ as it is from $(5, 0)$.

   a. $(10, 0)$ **b.** $(-6, 8)$ **c.** $(8, 6)$ **d.** (x, y)

6. Two positive numbers multiply to 49. What is the largest and smallest sum possible?

Neat Stuff

7. Here are the graphs of a unit circle and a line with slope m, intersecting at $(0, -1)$.

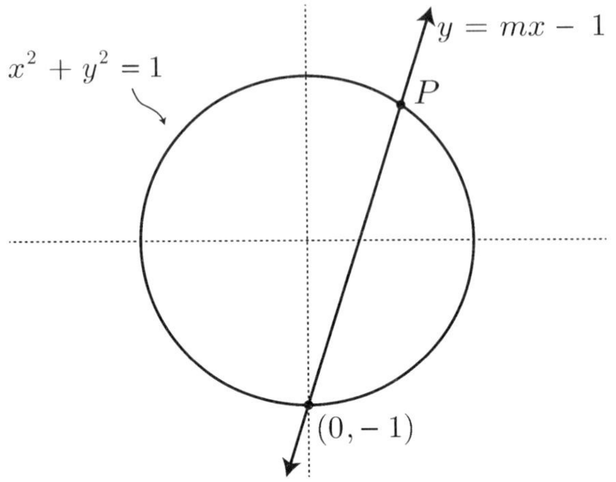

For each value of m given, determine the exact coordinates of point P.

a. 2 **b.** $\frac{3}{2}$ **c.** $\frac{4}{3}$ **d.** 4 **e.** 10 **f.** $\frac{a}{b}$

8. Let $f(x) = x^3 - 6x^2 + 4x + 8$. For each quadratic below, find the remainder when $f(x)$ is divided by the quadratic, then plot $f(x)$ and the remainder on the same axes (using techmology). What do you notice?

a. $(x-2)(x-6) = x^2 - 8x + 12$
b. $(x-2)(x-5) = x^2 - 7x + 10$
c. $(x-2)(x-4) = x^2 - 6x + 8$
d. $(x-2)(x-3) = x^2 - 5x + 6$

9. Let $f(x) = x^3 - 6x^2 + 4x + 8$. Do what you did on the last problem. What do you notice?

a. $(x-2)^2$ **b.** $(x-3)^2$ **c.** $(x-4)^2$ **d.** $(x-5)^2$

10. Find the equation of the tangent line to $f(x) = x^4$ at $x = 1$.

11. If you draw the altitudes from any interior point to all sides of an equilateral triangle, the altitudes' lengths add up to a constant. What about these shapes? For those that work, prove it; for those that don't, explain why they don't work.

a. a square
b. a rectangle
c. a rhombus
d. a parallelogram
e. a regular pentagon

f. an equilateral hexagon (not regular)
g. an equiangular hexagon (not regular)

12. The sum of two numbers is s and the product is p. Find the sum of the . . .

a. squares of the two numbers.
b. cubes of the two numbers.
c. fourth powers of the two numbers.
d. . . . a generalization?

Tough Stuff

13. Take a triangle, and move its points according to the rule $(x, y) \mapsto (ax + by, cx + dy)$. Find integer values for a, b, c, and d so that the new shape has a smaller area than the original shape, but still *some* area.

14. So we've discovered that this 120° point gives the least possible total distance to the three vertices. But what about other points? They're worse, but some are not much worse. Indeed, the shape of the points that are *equally bad* is interesting. What's it look like? What's it look like if you move outside the original triangle?

15. Find several triangles that have integer side lengths (with no common factors) and a 120° angle. Generalize?

16. Find some Matsuura triangles, or prove they do not exist. (See Problem Set 2 for the definition.)

17. Find this sum exactly:

$$0 + \frac{1}{100} + \frac{4}{10000} + \frac{9}{1000000} + \cdots + \frac{n^2}{10^{2n}} + \cdots$$

18. Find all integer solutions to this system of equations:

Nope, there's more.

$$a + b = cd$$
$$c + d = ab$$

Problem Set 5

Important Stuff

1. Two positive numbers multiply to 81. What is the largest and smallest sum possible?

Opener

Start a new sketch with your geometry software and follow these steps.

- Draw a circle.
- Draw a line that intersects the circle using two points that are outside the circle, and on opposite sides. Label one of these outside points A.
- Select the line and the circle, then click on "Construct" and "Intersection."
- Label the constructed points P and Q.

This step should bring *two* new points into the diagram.

1. Find the location of Q that maximizes the sum $AP + AQ$. You can drag the other unlabeled point on the line to move P and Q.
2. Find the location of Q that maximizes the product $AP \cdot AQ$.
3. Find the location of Q that minimizes the sum $AP + AQ$.
4. Move point A so that it is inside the circle. For this new location of A, find the location of Q that maximizes the product $AP \cdot AQ$.

---

2. Find a rectangle similar to a 7 by 11 rectangle and has the same numerical value for area and perimeter.

3. Todd starts with a 5 by 6 by 20 box.

a. Betul finds a box similar to Todd's box whose surface area and volume have the same numeric value. What are the dimensions of Betul's box?

b. Call the dimensions of Betul's box T, O, and D. Find the exact value of

$$\frac{1}{T} + \frac{1}{O} + \frac{1}{D}$$

4. For each of these points, determine whether it is three times as far from $(9,0)$ as it is from $(1,0)$.

   a. $(0,3)$ **b.** $(7,0)$ **c.** $(-2,2)$
   d. $(-3,0)$ **e.** $(1,-5)$ **f.** (x,y)

5. Do the "SAM" problem (Problem 6 from Set 2) if you haven't already.

6. Let $s = 2 + i$ and $m = 4 + 6i$ be complex numbers. Find each of these.

   a. $s + m$ **b.** $s - m$
   c. $3s + 3m$ **d.** sm (the product)
   e. is **f.** im

When you see i^2, make it -1. The number i is the imaginary square root of -1.

7. Simplify these seemingly nasty-looking expressions that involve square roots of square roots of negative numbers.

   a. $\sqrt{(3+4i)(3-4i)}$ **b.** $\sqrt{(5-12i)(5+12i)}$
   c. $\sqrt{(15+8i)(15-8i)}$ **d.** $\sqrt{(x+yi)(x-yi)}$

8. How far is each of these points from the origin?

   a. $(3,4)$ **b.** $(5,-12)$ **c.** $(15,8)$ **d.** (x,y)

9. **a.** Draw a graph of all the points (x,y) that are 5 units from the origin.
   b. Write an equation for the graph you just drew.

10. Find all 12 complex numbers $a + bi$ with integers a, b so that

$$\sqrt{(a+bi)(a-bi)} = 5$$

A complex number with integers a, b is called a *Gaussian integer*. You can look up one of the 12 answers.

Neat Stuff

11. For each point P, determine if it is twice as far from the *line* $x = 2$ as it is from the point $(-1,3)$.

   a. $(5,3)$ **b.** $(0,3)$ **c.** $(-2,4)$
   d. $(-3,2)$ **e.** $(-2,b)$ **f.** (x,y)

Is $x = 2$ vertical or horizontal? Decide by asking what points are on its graph.

12. Square ZACK has one vertex at $A(1,0)$. The center of the square is the origin $O(0,0)$. Find the coordinates of the other vertices.

13. Find all four solutions to the equation $x^4 - 1 = 0$.

Here's a useless reminder that $a^2 - 9 = (a+3)(a-3)$. Or is it?

14. Equilateral triangle DEB has one vertex at $B(1,0)$. The center of the triangle is the origin $O(0,0)$. Find the coordinates of the other vertices.

15. Find all three solutions to the equation $x^3 - 1 = 0$.

You can factor something by using long division. Divide, and if the remainder is zero, you've got two factors.

16. Regular hexagon BUSHRA has one vertex at $A(1,0)$. The center of the hexagon is the origin $O(0,0)$. Find the coordinates of the other vertices.

17. Find all six solutions to the equation $x^6 - 1 = 0$.

18. Let $z = \frac{\sqrt{2}}{2} + \frac{\sqrt{2}}{2}i$. Take powers of z until it becomes awesome, then tell us the value of z^{101}.

19. Find a complex number z so that $z^{12} = 1$ but no smaller positive integer n has $z^n = 1$. Did we say "a"? We meant "all" of them.

20. Let P be a point at $(0,-1)$ and A be a point inside the circle $x^2 + y^2 = 1$. Let the line AP intersect the unit circle at point Q.

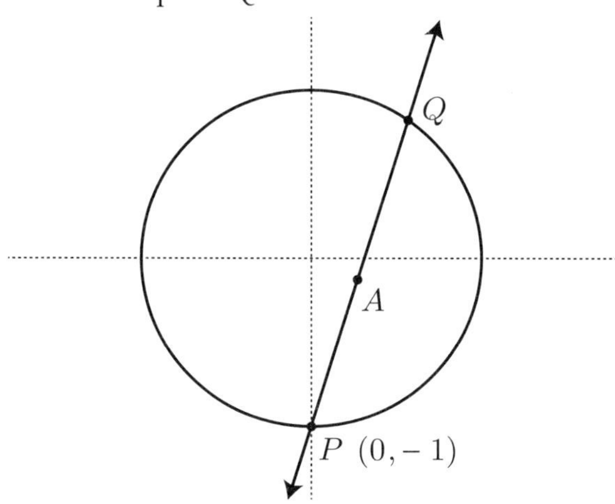

Calculate $f(A) = AP \cdot AQ$ for each of these points by determining the lengths AP and AQ.

a. $A = (\frac{1}{2}, 0)$ **b.** $A = (\frac{1}{4}, 0)$ **c.** $A = (-\frac{1}{2}, 0)$
d. $A = (\frac{1}{3}, -\frac{1}{2})$ **e.** $A = (\frac{1}{3}, \frac{1}{2})$ **f.** $A = (x, y)$

What happens if A is outside the circle?

Tough Stuff

21. Suppose n is a positive integer. Find a rule that determines whether there is a right triangle with area n and rational numbers as side lengths.

Problem Set 6

Opener

Graph the circle $x^2+y^2 = 65$ and determine all of its lattice points. A *lattice point* has integer coordinates.

Important Stuff

1. Solve for h and x.

$$h^2 + x^2 = 13^2$$
$$h^2 + (15 - x)^2 = 14^2$$

2. Find the perimeter and area of this triangle.

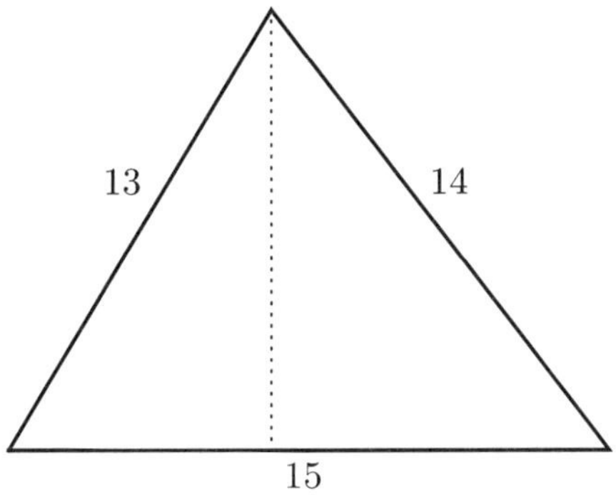

Don't use anything with the letter s when solving this problem.
The 13-14-15 triangle is called *Superheronian*. Its side lengths are consecutive integers and its area is an integer.

3. Find the side lengths of a triangle similar to the 13-14-15 triangle whose perimeter and area are equal numerically.

4. Find the area of this triangle.

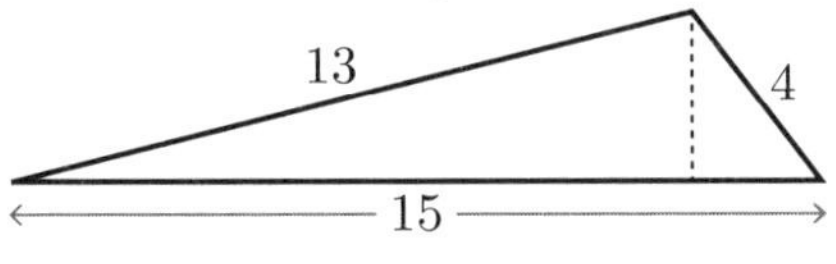

5. The *conjugate* of the complex number $z = a + bi$ is $\overline{z} = a - bi$.
   - **a.** If $z = 5 + 2i$, what is $\overline{z}$?
   - **b.** Let $w = 3 - 4i$. Calculate $w + \overline{w}$ and $w\overline{w}$.
   - **c.** Find a complex number v so that $v + \overline{v} = 14$.
   - **d.** Find a complex number v so that $v\overline{v} = 65$.
   - **e.** Find a complex number v so that $v + \overline{v} = 14$ *and* $v\overline{v} = 65$.

By the way, $w\overline{w}$ is just w multiplied by $\overline{w}$.

6. Find two numbers whose sum is 14 and product is 65.

7. The *magnitude* of the complex number $z = a + bi$ is $|z| = \sqrt{z\bar{z}}$.

   a. Find the magnitude of $z = 5 + 2i$ and of $w = 3 - 4i$.
   b. Rewrite the equation $|z| = \sqrt{(a + bi)(a - bi)}$ as something that doesn't have i in it.
   c. Can the magnitude of a complex number ever be zero? Can it be negative?
   d. Find a complex number whose magnitude is $\sqrt{65}$.

8. How many complex numbers $a + bi$ have integer a, b and magnitude $\sqrt{65}$?

9. a. Find the magnitude of $5 + 2i$.
   b. Find the magnitude of $(5 + 2i)^2$.
   c. Find a Pythagorean triple with hypotenuse 29.

10. Find a Pythagorean triple with hypotenuse 65.

11. Three positive integers add up to 25 and multiply to 360. What are the numbers?

360 is really divisible! Wouldn't it be awful if there were multiple answers here? Just awful. Seriously.

12. Find the volume and surface area of the box with dimensions $\frac{1}{15}$, $\frac{1}{6}$ and $\frac{1}{4}$.

Neat Stuff

13. Consider the transformation rule

$$(x, y) \mapsto (5x - 2y, 2x + 5y)$$

For each point, calculate the new point created by this transformation rule.

a. $P(1, 0)$ **b.** $Q(5, 2)$ **c.** $R(21, 20)$ **d.** $S(65, 142)$

14. The notation $|z|$ that is used for magnitude is the same symbol as the notation used for absolute value. Do real numbers have the same magnitude as their absolute value?

Real numbers *are* complex numbers, they're just numbers like $7 + 0i$ or $-\sqrt{2} + 0i$.

15. A *primitive Pythagorean triple* is a set of three positive integers a, b, c with $a^2 + b^2 = c^2$ where a, b, c share no common factors greater than 1.

   a. Find a primitive triple with hypotenuse 65.
   b. Find *both* primitive triples with hypotenuse 85.

6, 8, 10 is *not* one of these, nor is 60, 80, 100.

16. Find a way to make triangles that have integer side lengths and integer area, and make a bunch of them. *No formulas!*

17. Find more Superheronian triangles. See if you can find them all! There's a really small one, but 13-14-15 is the second-smallest.

Tough Stuff

18. Find all the triangles with integer-length sides whose area and perimeter have the same numerical value. Do they have anything else in common?

19. Find a number n that is the hypotenuse of exactly *four* primitive Pythagorean triples.

20. Same as the last one, but now exactly *eight* primitive Pythagorean triples!

21. Describe what types of positive integers n can be written in the form $n = a^2 + b^2$ for integers a, b. For example, 450 can be written this way: $450 = 21^2 + 3^2$.

22. The number 450 has eight odd factors: $1, 3, 5, 9, 15, 25, 45, 75, 225$. Odd factors can be split into factors in the form $4k + 1$ and $4k + 3$. 450 has 6 odd $4k + 1$ factors and 2 odd $4k+3$ factors. It turns out there's a remarkable connection between the number of these types of factors and the *number of different ways* 450 (or any positive integer) can be written in the form $n = a^2 + b^2$ for integers a, b. 450 can be written many different ways ($a = 3, b = 21$ is one of them). Figure out what the rule is—then prove that it works (harder). Don't forget that a or b can be negative or zero.

Problem Set 7

Opener

Take a sturdy triangle and suspend it inside a box (atop a can or cup). Now pour a huge amount of salt on it. What happens?

Triangles made of plastic or cut from cardboard are good. You will need a LOT of fine-grained salt! Boxes used to hold paper work well to catch the salt that drops.

This Sketchpad file shows a triangle after a severe pouring:
`http://www.tinyurl.com/saltgsp`

- What is so special about where the top of the salt heap is located?
- What is so special about where the ridges are located?
- Why is the top of the salt heap the same ________ to the three ________?

Important Stuff

1. Find the area of this triangle.

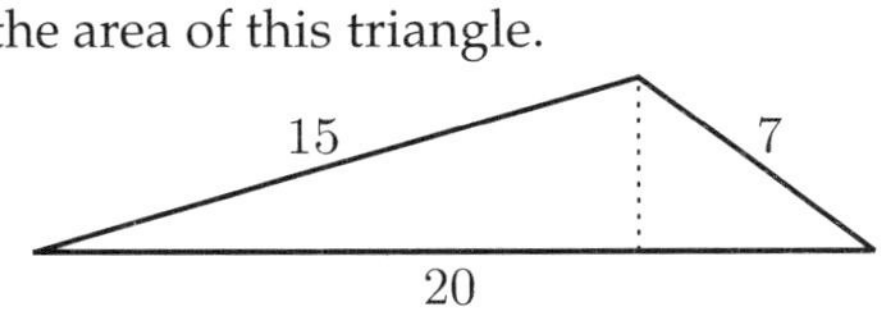

2. Find the area of this triangle.

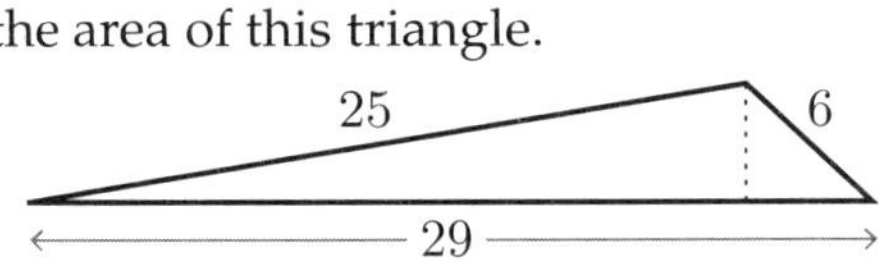

3. This file shows the triangle from Problem 1 above. `http://www.tinyurl.com/circlegsp` Adjust the included circle until it is the largest one that doesn't pass outside the triangle. What is the radius of this circle?

4. **a.** In your diagram from Problem 3, find a triangle with area 20. It might be hiding!

A look back at Problem Set 4 might be helpful.

b. Find an expression for the area of the entire triangle in terms of things that aren't the height of the triangle.

5. Consider the complex numbers $s = 2+i$, $a = 4+i$, $m = 4+6i$.

Plot the complex number $x + yi$ at the point (x, y) in the *complex plane*.

a. Plot and label s, a, and m in a complex plane.

b. Multiply each number by i, then plot and label each of the new numbers in the *same* complex plane.

c. Multiply each of s, a, and m by i *twice*, then plot and label each of the new numbers in the same plane.

d. Three times? Four times? Five times? Thirteen times? 101 times?

This all seems so familiar somehow . . .

6. Let $z = 1 + i$. Plot each of these in the same complex plane, and find the magnitude of each.

   a. z **b.** z^2 **c.** z^3 **d.** z^4 **e.** z^5

Remember, the magnitude of a complex number z is defined as $|z| = \sqrt{z\bar{z}}$. There may be other ways to interpret it . . .

7. Craig stands at the origin $(0,0)$ and stares at the powers of $1 + i$ as they are built. Describe what happens to the powers from his perspective: where do they go? how far away?

8. Let $z = 3 + 2i$. Find the *magnitude* of each of these complex numbers.

   a. z **b.** z^2 **c.** z^3 **d.** z^4

9. Let $z = \frac{3}{5} + \frac{4}{5}i$. Plot and label each of these on the same complex plane. Approximations are fine here.

   a. z **b.** z^2 **c.** z^3 **d.** z^4 **e.** z^5

10. Mahen stands at the origin $(0,0)$ and stares at the powers of $\frac{3}{5} + \frac{4}{5}i$ as they are built. Describe what happens as accurately as you can. How is it similar to what happens with the powers of $1 + i$? How is it different?

Neat Stuff

11. If $a = 3 + 4i$ and $b = 5 + 12i$, find the magnitude of a, the magnitude of b, the magnitude of ab.

12. **a.** Find the magnitude of $w = 2 + i$.

    b. Perform an operation to w that results in a complex number that has magnitude 5.

Yes, we mean 5, not $\sqrt{5}$.

13. Do stuff to take each of these complex numbers and produce a primitive Pythagorean triple.

    a. $4 + i$ **b.** $8 + 3i$ **c.** $15 + 4i$

    d. $16 + 7i$ **e.** $23 + 2i$ **f.** $42 + 9i$

14. **a.** Three numbers add up to 14 and multiply to 72. Find both sets of three positive integers that work here.

    b. Use the results to find two boxes whose surface area and volume equal one another.

15. The magnitude of $3 + 4i$ is 5. Find *all* the Pythagorean triples with hypotenuse 125 (including non-primitive ones).

16. Use the concepts from Problem 11 to find a primitive Pythagorean triple with hypotenuse 1105. Then another one! How many *are* there??

What's so special about 1105 anyway?

Tough Stuff

17. **a.** Find a way to generate all of the Pythagorean triples in which *the two leg lengths* are one away from each other. One example is 21, 20, 29.

b. Find a way to generate Pythagorean triples that are incredibly close to being 30-60-90 right triangles. They can't be, but they can be super close!

18. Besides being Heronian, this triangle has some other interesting feature. Find a way to generate more that have its special property.

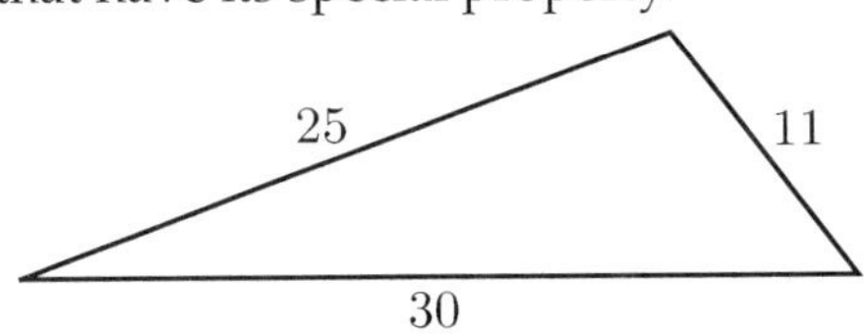

19. $\frac{A}{BC} + \frac{D}{EF} + \frac{G}{HI} = 1$. Here, each letter is a unique and distinct number from 1 to 9, and the denominators are two-digit numbers. Find a solution!

20. As you keep taking powers of $z = \frac{3}{5} + \frac{4}{5}i$, will they eventually wrap around onto themselves? In other words, are there powers k and m with $z^k = z^m$ for this z?

21. In triangle ABC, angle A is twice angle B, angle C is obtuse, and the three side lengths a, b, and c are integers. Determine, with proof, the minimum possible perimeter.

Problem Set 8

Opener

For each of these complex numbers, plot the number and its square in the same complex plane. You might want to draw segments from the origin to each number. If you need to measure lengths or angles, consider using geometry software.

Reminder: all activities here can be completed without formulas and without direct instruction.

$$A = 3 + i$$
$$S = 1 + 2i$$
$$H = 3 + 2i$$
$$L = \frac{\sqrt{3}}{2} + \frac{1}{2}i$$
$$I = i$$

You can type "3^0.5" to get $\sqrt{3}$ in geometry software.

Keep doing examples until you can describe *precisely* where the square is located in relation to the original number.

Extension questions: what about cubes? square roots? reciprocals?

Important Stuff

1. Let $z = \frac{12}{13} + \frac{5}{13}i$. Find and plot all of these on the same complex plane. (Estimate to three decimal places if you like.)

   a. z^0 **b.** z^1 **c.** z^2 **d.** z^3 **e.** z^4 **f.** z^5

2. Use geometry software to find the *angle* that forms when you go from one power of $z = \frac{12}{13} + \frac{5}{13}i$ to the next.

   You mean it's the same angle each time? Maybe, maybe not!

3. What does it look like in the complex plane when you add two complex numbers w and z? Give some examples. What does $w + w$ look like?

4. If a complex number w is multiplied by a real number c (also called a *scalar*), what happens to the magnitude? the direction? What if c is negative?

5. If a complex number w is multiplied by i, what happens to the magnitude? the direction?

6. Let $w = 1 + i$ and $z = \sqrt{3} + i$. Find the magnitude and direction of wz, and compare the results to the magnitude and direction of w and z.

7. What, in general, happens to magnitude and direction when you multiply two complex numbers?

8. **a.** Find the magnitude of $3 + 2i$.
   b. Calculate $(3 + 2i)^2$ and find its magnitude.
   c. What Pythagorean triple has hypotenuse 13?

9. Make some huge Pythagorean triples. Wow!

Neat Stuff

10. **a.** Three numbers add to 13 and multiply to 36. Find both sets of positive integers.
    b. Find two distinct rectangular boxes with volume 162 and surface area 234.
    c. If each box's dimensions are a by b by c, compute the product

$$(x - a)(x - b)(x - c)$$

and discuss any findings.

11. **a.** Point Q has coordinates $(3, 0)$ and point P has coordinates $(x, 0)$. Define $f(P)$ to be the square of the distance from point P to point Q. What does the graph of $f(P)$ look like?
    b. Point R has coordinates $(17, 0)$. Now define $f(P)$ as the sum of the squares of the distances PQ and PR. What does the graph of $f(P)$ look like?
    c. Point S has coordinates $(1, 0)$. You take it from here.
    d. If $f(P)$ is the sum of the squares of the distances PQ, PR and PS, what P produces the minimum of $f(P)$?

12. For each point P, determine if it is twice as far from the point $(-1, 4)$ as it is from the *line* $y = 1$.

a. $(-1, 2)$	**b.** $(-1, -2)$	**c.** $(23, 14)$
d. $(23, -14)$	**e.** $(-91, 52)$	**f.** (x, y)

13. Find an equation for all the points *three times* as far from $(-1, 4)$ as from the line $y = 1$.

14. Find an equation for all the points *half* as far from $(-1, 4)$ as from the line $y = 1$.

15. Find a triangle similar to the one below that has the same numerical value for its area and perimeter.

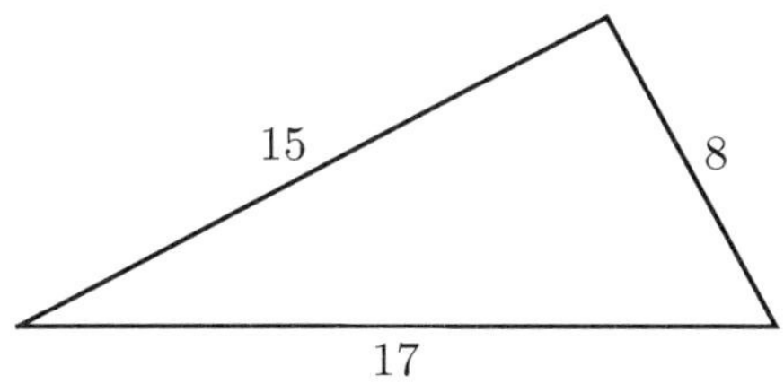

16. Find two triangles that have the same area and the same perimeter but aren't the same shape.

17. A right triangle has leg lengths a and b. Find the radius of its incircle.

Try it with numbers first!

18. Use the results from Problems 3–5 to prove that when you multiply a complex number by $a + bi$, it . . . uhh . . . what does it do?

19. Find a primitive Pythagorean triple whose hypotenuse length is 13^3.

20. Show how squaring the complex number $m + ni$ can be used to generate this identity that can be used to generate Pythagorean triples:

$$(m^2 - n^2)^2 + \left(2mn\right)^2 = (m^2 + n^2)^2$$

21. Let a be a complex number with magnitude 5, and b be a complex number with magnitude 13. Consider $a + b$: what could its magnitude be? Do you add magnitudes when you add complex numbers, or what?

22. Suppose x can be written as the sum of two squares, and y can also be written as the sum of two squares. Prove that xy can *also* be written as the sum of two squares.

23. Suppose a regular n-gon is placed with its center at the origin and one of its vertices at $(1, 0)$.

a. For what values of n will $(-1, 0)$ be a vertex of the n-gon?

b. For what values of n will $(0, -1)$ be a vertex of the n-gon?

c. Find all points that are on the regular 12-gon but aren't on any smaller n-gon.

24. Take all the vertices on the regular 12-gon as points in the complex plane. Square them and plot the squares. What happens?

25. You've now seen that squaring complex numbers generates Pythagorean triples. But does this squaring method generate *all* the Pythagorean triples? Explain.

Tough Stuff

26. Prove that at least one number in every Pythagorean triple must be even. Then prove that *exactly one* number in every primitive Pythagorean triple must be even.

27. Build an identity for generating primitive Pythagorean triples where the hypotenuse is a perfect square.

28. Hey, it's another division problem where we've provided one of the digits, an 8. The other digits are unknown.

```
                X X X X X
       X X X ) X X X X X X X X
               X X X
               -----
                 X X X X
                   X X X
                 -------
                     X X X X
                     8 X X X
                     -------
```

29. Sketch the graph of the equation $x^y = y^x$ in the first quadrant. What point is that, and why?

You'll know what point we mean once you get the graph going.

Problem Set 9

Opener

Let v be a complex number with magnitude 2.

- Draw a shape to indicate where v could lie in the complex plane.
- Pick a value of v and square it: where does it go? Describe all the possible places where v^2 could lie.

This Sketchpad file can help:
`http://tinyurl.com/complexgsp`

Important Stuff

1. Use complex numbers to make three ridiculously large, primitive Pythagorean triples.

2. Find the area of this triangle. Keep careful track of your steps and do not use any "instant winner" formulas.

While it is possible to "eyeball" the triangle by picking just the right altitude, please additionally work through the calculations. They'll be helpful later in this problem set.

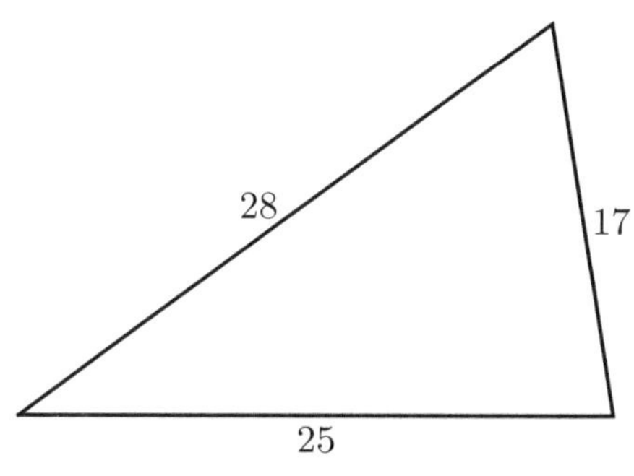

3. Explain why the area of a triangle is given by

$$A = \frac{1}{2}Pr$$

where P is the triangle's perimeter and r is the radius of its incircle.

4. Find the incircle radius of the triangle used in Problem 2.

5. Find the area of this triangle by following the same steps you followed in Problem 2.

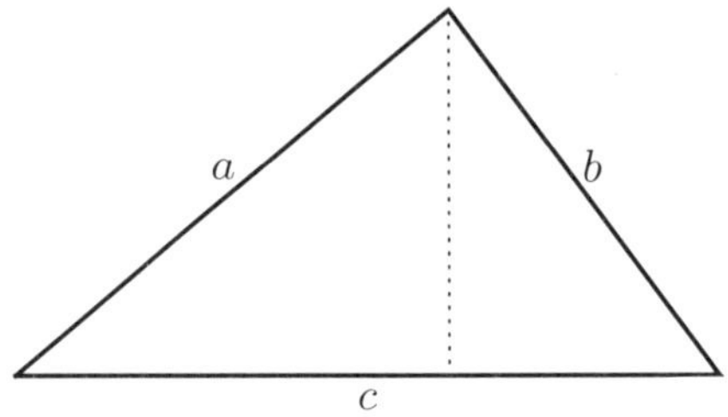

6. Find the incircle radius of a triangle whose side lengths are a, b, and c.

7. Take an 8-15-17 right triangle and a 5-12-13 right triangle and slap them together in some way to form a triangle with integer side lengths and integer area. You may need to scale one or both triangles first.

Triangles with integer side lengths and integer area are called *Heronian triangles*.

8. Find at least two more ways to turn the 8-15-17 and the 5-12-13 into Heronian triangles.

9. Use your ridiculous Pythagorean triples to make a few ridiculous Heronian triangles.

10. The complex numbers z and w are plotted below. Plot $z + w$ and indicate how you know where it should be.

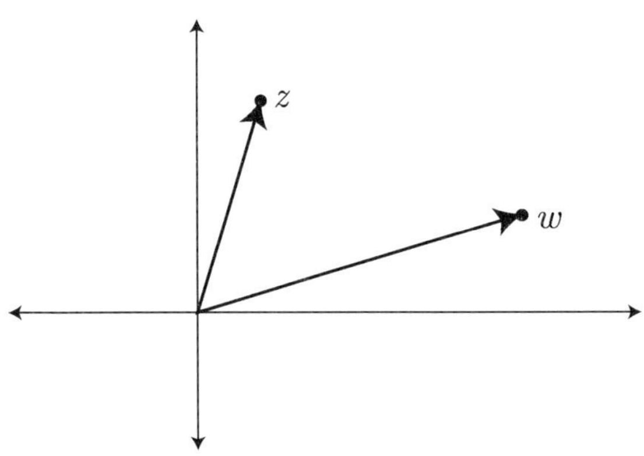

Neat Stuff

11. Compute $(1+2i)(1-2i)$. Here, $1-2i$ is the conjugate of $1+2i$.

12. Find the magnitude and direction of $1+2i$.

Geometry software can help you measure angles; remember, the "direction" angle is measured counterclockwise from the positive real axis.

13. Here are three complex numbers: $j = 3 + i$, $m = 3 + 2i$, $i = i$. They can be connected to make a triangle.

a. Plot these three numbers in the complex plane.
b. Multiply each number by $1+2i$, then plot the results in the complex plane.
c. Describe, as precisely as possible, how the new points' locations compare to the old.
d. How does the area of the new triangle relate to the area of the old triangle?

14. Find the surface area and volume of a 4 by 10 by 15 box and a 5 by 6 by 20 box.

15. Multiply these out and see what happens.

a. $(x-4)(x-10)(x-15)$
b. $(x-5)(x-6)(x-20)$
c. $(4x-1)(10x-1)(15x-1)$
d. $(x-\frac{1}{4})(x-\frac{1}{10})(x-\frac{1}{15})$

16. Find the dimensions of a box with the same total edge length and volume as the box with dimensions $\frac{1}{4}$ by $\frac{1}{10}$ by $\frac{1}{15}$.

17. So now you've got a cool formula for the area of a triangle given its sides. But some people bust out this crazy formula:

This thing s is called the *semiperimeter*, which is half the perimeter of the triangle.

$$A = \sqrt{s(s-a)(s-b)(s-c)}$$

Can you turn your cool formula into this one with an s?

18. Find a different Heronian triangle with the same perimeter and area as the 17-25-28 triangle.

19. A triangle has side lengths 15, 7, and x, and its area and perimeter have the same numerical value. Find all possible values of x.

20. Find an equation for all the points that are α times as far from $(-1, 4)$ as from the line $y = 1$. (Let α be a positive number.) Describe how the shape created depends on the value of α.

Make sure what you find agrees with what you learned through Problems 12–14 from Problem Set 8.

21. Suppose z is a complex number with magnitude 1 and direction θ. Then $z = a + bi$ with $a = \cos\theta$ and $b = \sin\theta$.

   a. Calculate z^2 directly by squaring $z = a + bi$. Wow!!

   b. Find a formula for $\cos 3\theta$.

If you're not working with trigonometry regularly, these last four will be relatively boring problems and are easily skipped.

22. Suppose z and w have magnitude 1, z has direction α and w has direction β. Let $z = a + bi$ and $w = c + di$.

   a. Calculate zw.

   b. What are the magnitude and direction of zw?

   c. What's the formula for $\sin(\alpha + \beta)$?

23. If $\tan A = a$ and $\tan B = b$, find a *simple* way to compute $\tan(A + B)$ without formulas.

Multiply! Find the right complex numbers and you're off to the races.

24. Pick angles A, B, and C that form a triangle. (You know what we mean.) Now let $\tan A = a$, $\tan B = b$, and $\tan C = c$. Which is bigger, the product abc or the sum $a + b + c$? Try another triangle and compare.

Tough Stuff

25. A triangle is uniquely determined by its side lengths, so it makes sense there is a formula for the area in terms of the three side lengths. Did you know that a triangle is also uniquely determined by its median lengths? That means there is a formula for the area in terms of the lengths of the three medians. Go find it and celebrate because it's awesome.

26. There is also a formula for the area of a triangle in terms of the lengths of its three altitudes. It is less awesome but still possible.

27. There is also a formula for the area in terms of the lengths of its three angle bisectors. Good luck with that.

28. Prove that every Pythagorean triple must have a multiple of 3, a multiple of 4, and a multiple of 5 in it somewhere.

Problem Set 10

Opener

Let v be a complex number with magnitude 2.

- Draw where v could lie in the complex plane.
- Pick a value of v and cube it: where does it go? Describe all the possible places where v^3 could lie.
- Where does $v + i$ go and how does it move as v changes? What about $v + 3$? $2iv$? $v^2 + v$? $v^2 - v$?

This Sketchpad file will be helpful:

`http://tinyurl.com/complexgsp`

"Plot Point" in Sketchpad locks a point's coordinates when you translate or zoom. Use this to fix the radius of your circle to the coordinate plane, since it keeps the correct radius if you change your scale.

Try experimenting with the "Trace" or "Locus" functions of your geometry software.

Important Stuff

1. Here are two Pythagorean triples: 5-12-13 and 85-132-157. Use them to produce more than one Heronian triangle. Then find its area and incircle radius using any method you like.

2. Solve these quadratic equations.

a. $(x-3)(x-17)=0$
b. $(2x-3)(2x-17)=0$
c. $(10x-3)(10x-17)=0$
d. $(100x-3)(100x-17)=0$

3. Solve these quadratic equations. Try to avoid using formulas.

a. $x^2-14x+45=0$
b. $4x^2-28x+45=0$
c. $100x^2-140x+45=0$
d. $10000x^2-1400x+45=0$

Hint: use the first one to help you do the second one!

4. **a.** Write a quadratic equation whose solutions are $x=\frac{2}{3}$ and $x=-\frac{4}{7}$.

b. Sorry, we meant one with no fractions, multiplied out, and with 0 by itself on one side. Try again.

c. Did you know that $\frac{2}{3}-\frac{4}{7}=\frac{2}{21}$? Interesting. Discuss.

d. In the quadratic equation $ax^2+bx+c=0$, what is the sum of the two solutions? What is the *average* of the two solutions?

5. Let $f(x) = ax^2 + bx + c$, and use the quadratic you used in the previous problem.

   a. Graph $f(x)$ using your favorite technology.
   b. What are the x-intercepts of the graph?
   c. What is the x-coordinate of the vertex?

The vertex is the turning spot.

6. What is the x-coordinate of the vertex of $f(x) = ax^2 + bx + c$ in terms of a, b, and maybe c?

7. Consider points $P(x, 0)$, $Q(3, 0)$, $R(17, 0)$, $S(1, 0)$, and $T(11, 0)$.

   a. Define $a(P)$ to be the distance PQ. What does the graph of $a(P)$ look like, and where is its minimum?
   b. Redefine $a(P)$ to be the sum of the distances PQ and PR. What does the graph of $a(P)$ look like, and where is its minimum achieved?
   c. Redefine $a(P)$ to be the sum of the distances PQ, PR, and PS. Again!
   d. Redefine $a(P)$ to be the sum of the distances PQ, PR, PS, and PT. Again!

8. Consider points $P(x, 0)$, $Q(3, 0)$, $R(17, 0)$, $S(1, 0)$, and $T(11, 0)$.

   a. Define $b(P)$ to be the *square* of the distance PQ. What does the graph of $b(P)$ look like, and where is its minimum achieved?
   b. Redefine $b(P)$ to be the *sum of the squares* of the distances PQ and PR. What does the graph of $b(P)$ look like, and where is its minimum achieved?
   c. Redefine $b(P)$ to be the sum of the squares of the distances PQ, PR, and PS. Again!
   d. Redefine $b(P)$ to be the sum of the squares of the distances PQ, PR, PS, and PT. Again!

9. Which of the following points is a total of 30 units away from $(-9, 0)$ and $(9, 0)$?

   a. $(-15, 0)$ **b.** $(0, 12)$ **c.** $(6, 11)$
   d. $(9, -\frac{48}{5})$ **e.** $(12, b)$ **f.** (x, y)

As before, these problems are not multiple choice. There may be more than one correct answer.

10. The distance from (x, y) to the point $(9, 0)$ is $\frac{3}{5}$ of its distance to the line $x = 25$. Which of the following points makes that statement true?

    a. $(15, 0)$ **b.** $(0, -12)$ **c.** $(13, 6)$
    d. $(9, \frac{48}{5})$ **e.** $(-12, b)$ **f.** (x, y)

Neat Stuff

11. The *eccentricity of an ellipse* is the ratio $\frac{c}{a}$ of two distances: the distance from the center of the ellipse to a focus (called c), and the distance from the center of the ellipse to a vertex (called a). For any ellipse discovered in Important Stuff, compute its eccentricity.

12. The distance from (x, y) to the point $(9, 0)$ is $\frac{4}{5}$ of its distance to the line $y = x$. Which of the following points makes that statement true?

a. $(7, 2)$ **b.** $(23, -2)$ **c.** $(11, -14)$
d. $(19, -19)$ **e.** $(27, -18)$ **f.** (x, y)

13. Show that this is true:

$$(a^2 + b^2)(c^2 + d^2) = (ac - bd)^2 + (ad + bc)^2$$

Squares, squares, everywhere.

14. Suppose m can be written as the sum of two squares, and n can also be written as the sum of two squares. Prove that mn can *also* be written as the sum of two squares.

15. Use the result in Problem 13 to prove that when two complex numbers are multiplied, their magnitudes are multiplied.

16. Suppose the sum and product of two complex numbers z and w are both real. Prove that either z and w are both real numbers, or $w = \bar{z}$.

17. Find the eccentricity of the shape in Problem 12.

18. Find two values of $\tan\alpha$ and $\tan\beta$ so that $\tan(\alpha + \beta) = 5$.

If you haven't yet, come up with a formula for $\tan(\alpha + \beta)$ in terms of $\tan\alpha$ and $\tan\beta$.

19. Calculate this:

$$\frac{(5+i)^4}{(239+i)}$$

Any guesses on what this might be useful for?

Useful?! Math is supposed to be useful?

Tough Stuff

20. Let S be the set of complex numbers z with magnitude 2. Find a function (such as $z \mapsto z^2$) that produces an ellipse (and *not* a circle) as the output when S is used as the input.

21. The complex number w is a member of set M if the rule $z \mapsto z^2 + w$ with starting point 0 never has $|z| > 2$.

We mean here that the rule is executed repeatedly. For example, if $w = i$ then the sequence of z values is $0, i, -1 + i, -i \ldots$ and the hope is that none of those has magnitude more than 2.

- **a.** Find some numbers w that are in set M, and some that aren't.
- **b.** Find a maximum bound for the *area* of the shape made by set M in the complex plane.
- **c.** What does set M look like if it is graphed in the complex plane?

22.
- **a.** Evaluate $\tan 89°$ to two decimal places.
- **b.** How many degrees are in 1 radian? Give your answer to three decimal places.
- **c.** What is going on here? Is this a coincidence?

Probably not!

Problem Set 11

Opener

Here is David's data.

x	y
1	1
3	8
5	3
7	8

Decide, without using anything other than your brain and perhaps some paper and pencil, on an equation for a line that "best" represents his data table. Be prepared to defend your choice: why did you pick this line above others?

A graph of the four data points can be found at the end of this problem set.

Important Stuff

1. One way to test the "badness" of a line toward data is to compute the *sum of squared errors*. Use the data from the box above.
   - **a.** Trevor used the line $y = x$ and wants you to confirm that the sum of squared errors is 30.
   - **b.** Marakina used $y = x - 1$ and Olimpia used $y = x + 1$. How'd they do?
   - **c.** Barb says there must be a rule, in terms of b, for the sum of squared errors of $y = x + b$. Go find it.
   - **d.** Kate screams "Oh snap, it's quadratic!" then finds the best possible b. Join her . . .

2. So, given a slope, it's possible to find one best line of that slope for the data (according to the sum of squared errors). Hop to it, and find the best line for each slope.

   a. slope 2　**b.** slope 3　**c.** slope 4
   d. slope 0　**e.** slope -1

Dividing up this work is fine, but make sure the answers are accurate!

3. Graph the data and all six of the lines from the last two problems on the same axes. Wow!!

4. Marcelle hands you this new data set.

x	y
-3	-4
-1	3
1	-2
3	3

In terms of m, find the sum of squared errors of $y = mx$. Then determine the best m possible.

5. Type the original four points into a piece of technology and tell it to compute a linear regression.

A graphing calculator might be best for this. Some geometry software will not compute the linear regression automatically.

a. What is the slope of the linear regression?
b. What important point does the graph of the linear regression pass through?
c. Compute the badness of the linear regression. Is it simply the "best"?

6. A second way to test the "badness" of a line toward data is to compute the *sum of absolute errors*. Use David's original data.

a. Lauren used the line $y = x$ and wants you to confirm that the sum of absolute errors is 8.
b. James used $y = x - 1$ and Timon used $y = x + 1$. How'd they do?
c. Allison tried $y = x + \frac{1}{2}$. How'd she do?
d. Becky says there must be a rule, in terms of b, for the sum of absolute errors of $y = x + b$. Go find it.
e. Sketch the graph of Becky's rule as a function of b. What do you notice?

Neat Stuff

7. In the Opener from Problem Set 10, you visualized the locus of all points $v^2 + v$ when the complex number v has magnitude 2. Now, check out this Sketchpad file:

`http://tinyurl.com/comppolygsp`

Use the sketch to show that $z = 2$ is a solution to the equation $z^2 + z = 6$. Then find *all* of the solutions to $z^2 + z = 6$, resizing the circle as necessary to track them down.

8. Use the file to find at least one solution to each of these equations.

- **a.** $z^2 - z = 6$
- **b.** $z^2 - z = 12$
- **c.** $z^2 - z = -9$ (do a quick approximation)
- **d.** $z^2 - z = 20$

Start by using the sliders to change the function, then set the green circle's radius to 3.

You're going to need a bigger circle.

9. Given z in the complex plane, how can you construct the conjugate $\overline{z}$? What's the conjugate of $z^2 + z$?

10. The distance from (x, y) to the point $(25, 0)$ is 30 more than its distance to $(-25, 0)$. What points (x, y) make this statement true?

If your answer includes $(15, 0)$, you're overlooking something . . .

11. The distance from (x, y) to the point $(25, 0)$ is $\frac{5}{3}$ of its distance to the line $x = 9$. Test some points, then find an equation!

12. The *eccentricity of a hyperbola* is the ratio $\frac{c}{a}$ of two distances: the distance from the center of the hyperbola to a focus (called c), and the distance from the center of the hyperbola to a vertex (called a). For any hyperbola discovered in this problem set, compute its eccentricity.

13. For David's data in the Opener, find the unique line with the smallest sum of absolute errors.

14. Brian hands you this list of points:

t	h
1	3
1	7
3	5
5	3
5	7

Find the line that has the smallest sum of absolute errors. Sorry, it's not unique: find them all.

15. Let $z = 34 + i$ and $w = 55 + i$. The product zw can be written as a scalar multiple of another complex number in the form (something) + i. What's the something? (Oh, what about $\frac{z}{w}$?)

34 and 55, you say? I'd be fibbing if I said I'd never seen *those* numbers before. Perhaps the answer is 89.

16. Let $a = 2 + i$, $b = 5 + i$, $c = 13 + i$, $d = 21 + i$. Find the direction of the product $abcd$. Can this be generalized?

17. Find the three solutions to $x^3 = i$ using any means, including a sketch.

18. **a.** Multiply out $(a + bi)^3$.
b. If z is a complex number with magnitude 1 and direction θ, what are its coordinates?
c. Write rules for $\cos 3\theta$ and $\sin 3\theta$ based on the first two parts of this problem.
d. The usual rule for $\cos 3\theta$ is $4\cos^3\theta - 3\cos\theta$. Does this agree with what you found?

If you're not a trig fan, just move along, nothing to see here.

19. **a.** Expand the expression $(x + 2)^3$.
b. How many faces are there on a cube? How many edges? How many vertices? Nifty.
c. Does this pattern continue at all, in lower or higher dimensions?

20. Suppose $\tan A = \frac{1}{x}$ and $\tan B = \frac{1}{y}$, where x and y are integers. Is it possible for $\tan(A + B)$ to also be in the form $\frac{1}{z}$ for some other integer z? If so, how?

21. Let z be a complex number with magnitude 1. If $z = x + yi$ and $y > 0$, explain why $y = \sqrt{1 - x^2}$.

22. Let $z = x + (\sqrt{1 - x^2})i$ as in the last problem.
a. Expand z^2 and look at the real part. Zeros?
b. Expand z^3 and look at the real part. Zeros? Connections?
c. Expand z^4 and look at the real part.
d. Graph the real part of z^2, z^3, z^4 as a function of x.
e. What might z^n look like?

Many computer algebra systems, including the TI-Nspire software, can handle this.

Tough Stuff

23. So, you managed to find two noncongruent boxes with the same surface area and volume? Fine, but can you find *three* noncongruent boxes with the same surface area and volume?

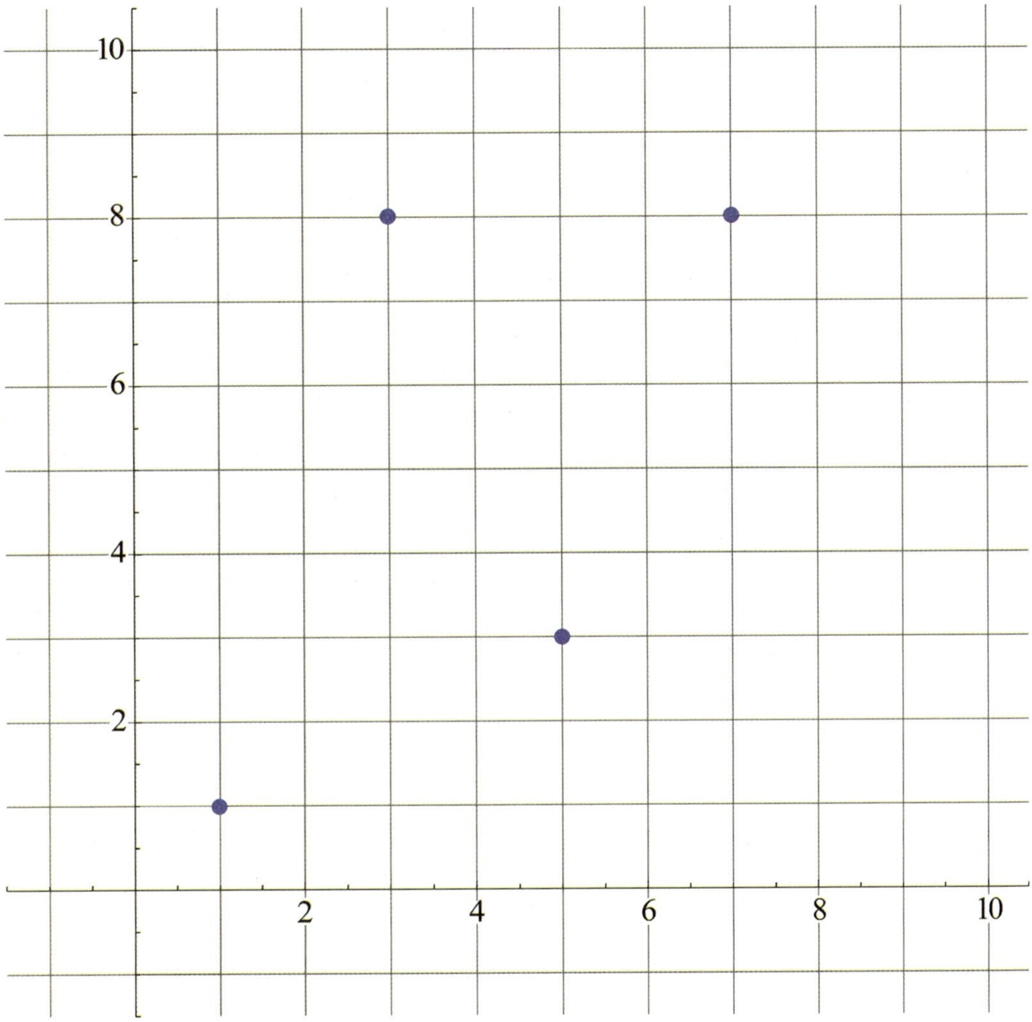
10
8
6
4
2
2
4
6
8
10

Problem Set 12

Opener

Sarah has this set of data giving n, the number of people enjoying Problem Set d.

At least it's not always decreasing.

d	n
2	40
3	30
5	35
6	20
9	10

Use what you learned from Problem Set 11 to find the line that minimizes the sum of squared errors. What "badness" did you get for this line? Check to see that other "close" lines have higher badness.

Important Stuff

1. **a.** Write an expression for the "badness" (sum of squared errors) for an arbitrary line $y = mx + b$ for the data above. *Don't expand any squared things in parentheses!* Make sure everyone at your table agrees on an answer.

   b. The crazy expression you made is equivalent to this:

   Check it if you want to. I don't recommend it!

$$5(b + 5m - 27)^2 + 30(m + 4)^2 + 100$$

   So what? What's this expression got going for it?

2. Use the badness expression

$$5(b + 5m - 27)^2 + 30(m + 4)^2 + 100$$

   to find one or two lines whose badness is exactly 130.

3. If A, B, and C are positive integers with $A \leqslant B \leqslant C$, find all possible solutions to

$$\frac{1}{A} + \frac{1}{B} + \frac{1}{C} = \frac{1}{2}$$

What is the largest possible value of A? Given A, what is the largest possible value of B? Keep at this until you exhaust all options.

4. Multiply out the equation in Problem 3 so there are no fractions. What do you notice?

5. Find all possible rectangular boxes with integer side lengths whose surface areas are numerically equal to their volumes.

Neat Stuff

6. The distance from (x, y) to the point $(1, 0)$ is α times its distance to the line $x = -1$. Figure out how the graph of this set of points depends on the number α.

7. The function $f(z) = z^3 + 4z$ is an *odd* function. What does an odd function do to the set of all complex numbers z with magnitude 2? Use this Sketchpad file to investigate:

`http://tinyurl.com/comppolygsp`

What about even functions?

8. Find or approximate the *three different* solutions to the equation

$$z^3 + z + 2 = 0$$

9. **a.** Use the sketch from Problem 8 to find the *five* complex numbers that make $z^5 = 1$.

b. Find the five complex numbers that make $z^5 = -1$.

10. Find all *ten* solutions to $z^{10} = 1$, and plot them in the complex plane.

If $z^{10} = 1$, what are the possible values of z^5?

11. Consider the function $w = f(z) = z^2 + 4z + 5$.

a. Mary uses a really small circle S as input. What does the output look like? By small we mean *really* small.

b. Tina likes large circles so she grabs Mary's circle and drags it until it's huge. What happens? Does the output ever cross the origin? How many times?

12. The badness expression in Problem 1 can also be written this way:

$$155\left(m + \frac{5}{31}b - \frac{111}{31}\right)^2 + \frac{30}{31}(b - 47)^2 + 100$$

Explain what this is good for, and see if you can find any similar ways to write the expression.

13. Compute the "badness" function for the data from the Opener in Problem Set 11 as a function of m and b. Then, rewrite that expression in a clever way to find the value of m and b that minimizes the badness.

Let A and B be two points. What is the locus of points P such that the ratio AP/PB is a fixed number r? Check out problems 14–17 if you're interested, otherwise skip 'em.

14. Let A and B be two points and let P be a point that is not collinear with A and B. There is a point X along the segment $\overline{AB}$ such that $AX/XB = AP/PB = r$. What is special about the location of X? Feel free to use geometry software to investigate this. To make GSP construct point X at the proper spot, measure AP and PB, then calculate the ratio $AP/(AP + PB)$. Double-click point A to mark it as the the center of a dilation, then dilate the point B using the marked ratio $AP/(AP + PB)$.

In Sketchpad, mark a center of dilation by double-clicking it.

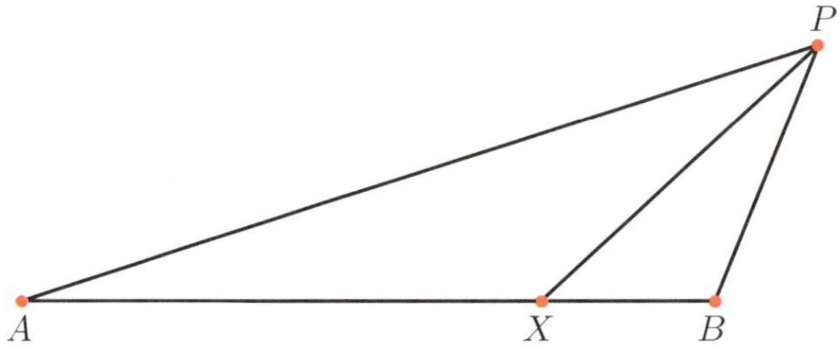

15. Prove what you found in Problem 14 by determining the ratio of the areas of triangles AXP and BXP in two different ways.

16. There's another point on the ray $\overrightarrow{AB}$ that is important. Let Y be a point on $\overrightarrow{AB}$ (besides X) for which $AY/YB = AP/PB = r$. What is special about the location of Y?

That dotted line is not a mistake . . .

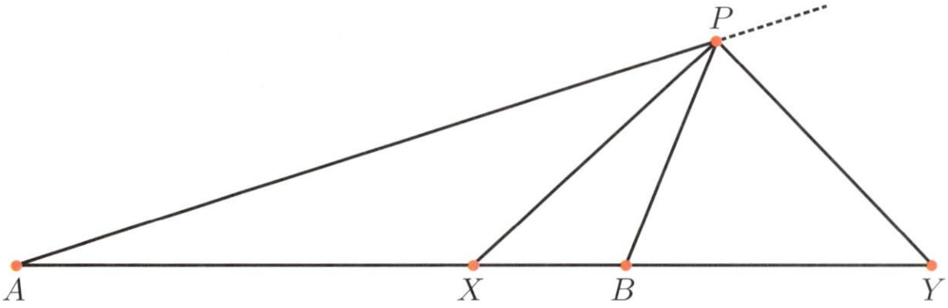

17. Determine the measure of $\angle XPY$ and its consequences for the possible locations of point P.

18. In radians, find the smallest positive x solving $\tan x = \frac{1}{5}$.

19. Let $f(x) = x - \frac{x^3}{3} + \frac{x^5}{5} - \frac{x^7}{7} + \cdots$. Values of $f(x)$ converge when x is between -1 and 1.

- **a.** Approximate $f(\frac{1}{5})$ to six decimal places. You might need to take a few terms.
- **b.** Approximate $f(\frac{1}{239})$ to six decimal places. Did this take more terms, or fewer terms?
- **c.** Evaluate $16f(\frac{1}{5}) - 4f(\frac{1}{239})$ to five decimal places. Wow!
- **d.** Find the direction of the complex number

$$\frac{(5+i)^4}{239+i}$$

What might this be useful for?

Tough Stuff

20. Here's a graph of a candidate fit line and the data from this problem set's Opener. Find the line that minimizes the total shaded area.

Early on, this method was proposed for determining the best fit line for data. It's not so good.

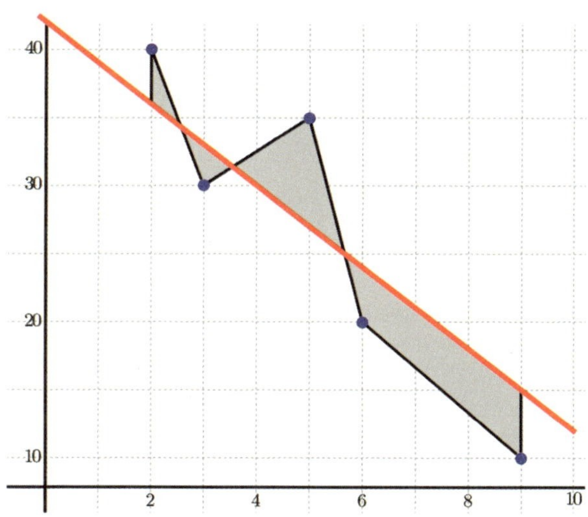

21. Factoring $x^5 - 1$ over the integers isn't too bad, it's

$$(x-1)(x^4 + x^3 + x^2 + x + 1)$$

The tough part is to factor that quartic piece into two quadratic factors, where each quadratic factor has real coefficients (not restricted to integers or fractions). Find the factoring, then look for connections between the factoring and the construction of a regular pentagon.

Problem Set 13

Opener

Karen puts 3 regular polygons together so they meet at a point with no overlap and no empty space. One option is to shove together 3 hexagons, but there are other ways.

Don't worry about whether these polygons will fill the entire plane this way. We're just looking at one vertex.

Find all the ways you can fit 3 regular polygons together at a point. Write each set in increasing order by the number of sides in the polygons.

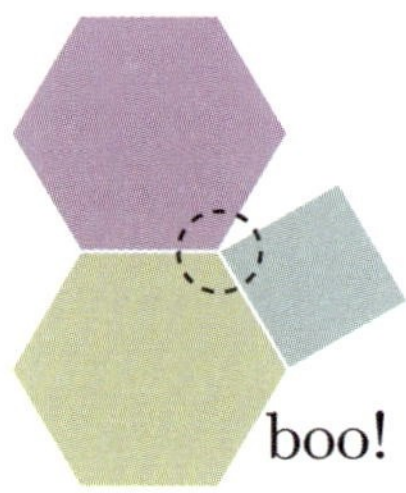

What! Wow. Why?

Important Stuff

1. Find all the ways you can fit 4 regular polygons together at a point.

2. Find all the positive integer solutions to this awesome equation, where $3 \leqslant G \leqslant A \leqslant B \leqslant E$:

$$\frac{1}{G} + \frac{1}{A} + \frac{1}{B} + \frac{1}{E} = 1$$

3. Find all the ways you can fit 5 regular polygons together at a point . . . 6 regular polygons . . . 7 regular polygons.

4. Pedro, Arden, Lindsey, and Sandy form a square. They just met and now they want to connect themselves. Connect them with paths so the total distance of all the paths is as small as possible. (Two people will be considered *connected* if there is any path from one to the other. Directly connecting is okay, but not required.)

The four people are friends so they will be *PALS*. But if they get in a fight, they *SLAP*.

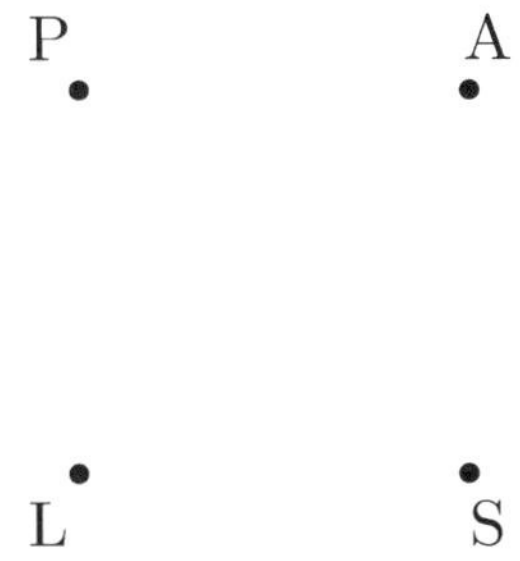

Neat Stuff

5. Find some right cylinders that have the same numerical value for volume and surface area.

The surface area of these right cylinders includes the top and bottom discs.

6. If Jocelyn tells you a right cylinder's total surface area and total volume, is that enough information to uniquely determine its dimensions?

7. Use this Sketchpad file:
`http://tinyurl.com/comppolygsp`
Investigate how the function $f(z) = z^2 - 2z + 4$ operates on different "magnitude circles" (circles centered at 0). Estimate with very poor accuracy the two solutions to $z^2 - 2z + 4 = 0$. (They are both on the same magnitude circle.)

8. Let $f(x) = x^3 + 6x + 3$. Within 30 seconds total, estimate $f(\frac{1}{100})$ and $f(100)$.

29 . . . 28 . . . 27 . . .

9. Consider the function $f(z) = z^3 + 6z + 3$.
   a. What does the output of $f(z)$ look like when you use a very small magnitude circle?
   b. What does the output of $f(z)$ look like when you use a very large magnitude circle?
   c. Explain how you know that there *must* be a magnitude circle that contains a point z with $f(z) = 0$.

This one has z instead of x, so it's clearly very different!

If you can't tell, your circle isn't small enough yet.

10. Let $f(x) = x^3 + 12x - 4$. Within 60 seconds, estimate each of the following.
    a. $f(\frac{1}{100})$ **b.** $f(-\frac{1}{100})$ **c.** $f(100)$ **d.** $f(-100)$

11. *Without using technology*, answer the same questions in Problem 9 but for $f(z) = z^3 + 12z - 4$.

12. Convince yourself and everyone else that this is true:

> If $f(z)$ is a polynomial of degree n then there must be at least one complex number z with $f(z) = 0$ in the complex plane.

"Degree n" just means that the polynomial starts $f(z) = az^n + \cdots$

13. Use exterior angles to explain why the stuff with the regular polygons just happened.

14. So the tiling of the two hexagons and the square from the box didn't work out. *Or did it!*

- **a.** Calculate $\frac{1}{6} + \frac{1}{6} + \frac{1}{4}$. How much more than $\frac{1}{2}$ is it?
- **b.** How many "missing" degrees are there at the vertex? How many "missing" degrees would there be for 24 vertices?
- **c.** Is there a tiling with two hexagons and a square at each vertex? What does it look like?

It won't tile the plane, but it might tile something else . . .

15. There's another option: tilings that overlap. For example, a tiling of three heptagons (7-sided figures). What might that look like, and is there any relationship to the fractions?

16. Use geometry software to construct the locus of points (x, y), whose distance to the point $(1, 0)$ is α times its distance to the line $x = -1$.

17. Consider the function $f(z) = z^2 - z$.

- **a.** Find a solution to $z^2 - z = 1$ using geometry software.
- **b.** Find a way to use this sketch to build a golden rectangle.

18. Use geometry software to approximate the five solutions to $x^5 + 2x^2 + 10 = 0$. How many of the roots are real numbers?

19. Use geometry software to approximate the four solutions to $x^4 + 3x^2 + 1 = 0$. Yes, there are four solutions and not two. What happens, in general, to even functions?

20. Find the four roots of $x^4 + 3x^2 + 1 = 0$ by factoring.

21. Track down the solutions to $x^3 - 7x^2 + 15x - 9 = 0$. What does a "double root" look like in the complex plane? How many solutions does this equation have?

Tough Stuff

22. Construct a regular pentagon using the quadratic factoring of $x^5 - 1$ from Problem 21 from Set 12, then look for quadratic factorings of other polynomials in the form $x^n - 1$.

23. Find some connections between what we've been doing with the tangent function and Taylor series.

24. Use the Taylor series for $\sin x$, $\cos x$, and e^x to show that $e^{i\pi} + 1 = 0$.

25. What is one possible value of i^i? (Use Google or a TI-Nspire handheld in radian mode.) How on earth would one arrive at such a thing? Why did we say "one possible value"?

Problem Set 14

Opener

Use this Sketchpad file:

`http://tinyurl.com/comppolygsp`

Investigate the function $w = f(z) = z^3 - 2z^2 - 5z + 6$.

- When we restrict the input z to a very small magnitude circle, what does the output look like?
- When we restrict the input z to a very large magnitude circle, what does the output look like?
- Explain how you know that there *must* be a magnitude circle that contains a point z with $f(z) = 0$.

Some Stuff

1. David says he can use the method of the Opener to show that, for *any* polynomial, there is at least one z with $f(z) = 0$. He's just nuts, right?

2. A tetrahedron has three of its faces meeting together at right angles. Find a relationship between the areas of its four faces.

3. How can you prove that the point inside a triangle that makes 120° angles with the vertices has the shortest total distance to the three vertices in a triangle? Does your proof work for *any* triangle?

4. Find two non-congruent triangles with integer side lengths that share the same area, A, and the same perimeter, P.

5. **a.** Let $z = 1 + i$. Graph several positive integer powers of it. What happens?
   b. Do the same thing, but start with $w = \frac{1}{2} + \frac{1}{2}i$.
   c. What is the value of the infinite sum $1 + w + w^2 + w^3 + \cdots$? Can you show this geometrically?

6. A more general version of the unit fraction problem is

$$\frac{1}{x_1} + \frac{1}{x_2} + \cdots + \frac{1}{x_n} = \frac{k}{2}$$

where n, k, and all x_i are positive integers. Find some conditions on n and k for which this problem has a solution.

7. Revisit Problem 4 from Set 13, but now suppose that the four friends are arranged in a rectangle rather than a square. How does the solution change? If they are arranged in a general quadrilateral, will the solution be unique? What if there are five points in an arbitrary arrangement?

8. Three regular polygons can meet at a point with no overlap and no empty space in several ways. But which ones tessellate the plane?

9. For each regular n-gon, find the side length that will make its area numerically equal to its perimeter.

a. 3-gon **b.** 4-gon **c.** 6-gon
d. 8-gon **e.** 12-gon **f.** n-gon

10. Given a specific surface area and volume, how can we find all the boxes with that surface area and volume? What if we are restricted to boxes with integer side lengths?

11. Can every Heronian triangle be deconstructed into two Pythagorean triples?

Remember, a *Heronian triangle* has integer side lengths and integer area.

12. In Problem Set 4 we intersected the unit circle $y^2 + x^2 = 1$ with a line of slope m passing through $(0, -1)$. The other intersection with the unit circle could be found in terms of m, and interesting things happened: we could find many rational-coordinate points on the unit circle.

Repeat this process with the unit *hyperbola* $y^2 - x^2 = 1$ with a line of slope m passing through $(0, -1)$. Find the other intersection in terms of m and enjoy the magic.

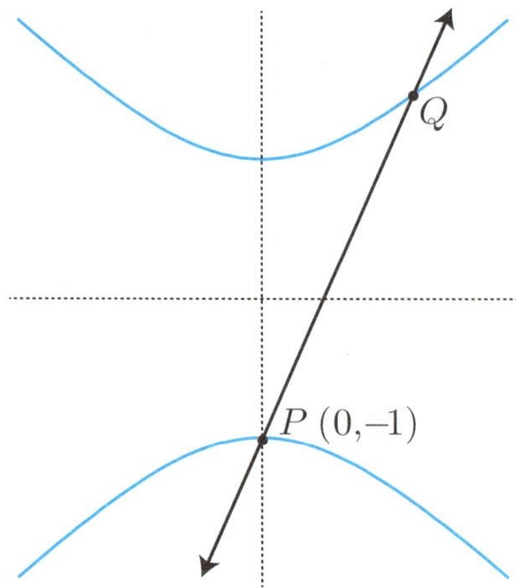

13. What is the square root of i? The cube root? The sixth root? How many answers are there??

14. Find another complex number whose direction is 60° counterclockwise from these given complex numbers, as seen from the origin.

a. $1+i$ **b.** $3-2i$ **c.** 5 **d.** $a+bi$

15. Find a third point in the coordinate plane that will complete an equilateral triangle with these given two points.

a. $(3,4)$ and $(4,5)$ **b.** $(2,1)$ and $(5,-1)$
c. $(3x,2)$ and $(3x+5,2)$ **d.** (x,y) and (w,z)

16. The "taxicab distance" (a.k.a. "Manhattan distance" or "rectilinear distance") between any two points in a plane is equal to the sum of the absolute differences of their coordinates. If you think of the lattice points on a graph as being the intersection of streets and avenues, the taxicab distance between two points represents how far a taxi must go to get from one intersection to another *travelling only on the streets and avenues* of the plane. The taxicab distance between $(3,5)$ and $(7,3)$ is calculated as $|7-3|+|5-3| = 4+2 = 6$. Note that there are several paths between the two points that have this minimum path length.

a. Jason is a fireman in brand new Gridville, which only has three houses at $(1,4)$, $(6,1)$, and $(9,5)$. Where should his firehouse be built to minimize the average taxicab distance to the three houses? This Sketchpad file may be useful:
`http://www.tinyurl.com/taxicabgsp`

b. Two more houses are built in Gridville at locations $(2,7)$ and $(3,6)$. Fortunately, Jason's firehouse is easily moved. Where should he move it?

c. Stop and reflect on where these "best" points are and how they relate to the locations of the houses. Notice anything?

d. A sixth house is built at $(6,6)$. Find a new firehouse point for these six houses. And more! How many best possible locations are there now?

e. Farmville crops up nearby with two neighborhoods of houses. The houses are located at $(2,8)$, $(2,9)$, $(3,7)$ and $(3,10)$, as well as at $(5,1)$, $(7,4)$, $(8,2)$, and $(10,5)$. Find all the firehouse spots for this set of houses.

17. Go back through the previous problem and find firehouse spots that have the lowest possible *maximum* taxicab distance to any house. How are these points related to the median?

18. There may or may not exist a special tetrahedron with integer side lengths, integer areas of faces, and integer volumes. Find one, or prove that they do not exist.

19. Is it possible for a clock to have 120 degrees between all three hands if it is

a. a continuously-moving second hand?

b. a clock with discrete second "ticks" (with minute and hour hands moving incrementally)?

20. Go back to Brian's data from Problem 14 on Set 11. Suppose $h = mt + b$ is an attempt at a best fit line and its badness is measured using the sum of absolute errors. In the m-b plane, what is the locus of all points (m, b) corresponding to the same badness?

21. Jemal asks you to close your eyes and imagine a four-dimensional hyper-box with dimensions A, B, C, and D. Find all possible dimensions for this hyper-box so that $A \leqslant B \leqslant C \leqslant D$ are integers, and its four-dimensional hypervolume has the same numerical value as its total "face-volumes."

CHAPTER 2 Facilitator Guide

These facilitator notes are designed to be used as needed. Each problem set has two components:

1. **Goals of the Problem Set:** here we lay out what the principal ideas of each problem set are.
2. **Notes on Selected Problems:** we identify a few problems that are worth going over in a whole group discussion.

We will put our emphasis on the main goals of each lesson, drawn from the problems in the "Important Stuff."

Problem Set 1

Goals of the Problem Set

This course is about representation and parameterization, and the problems here are intended to introduce some of the key questions and concepts that will be seen throughout.

A key problem is introduced: whether or not the surface area and volume of a box can uniquely define its dimensions. This problem will be revisited and refined a number of times through the course. A similar but more straightforward question is asked first: whether or not the perimeter and area of a rectangle uniquely define its dimensions. Other problems then introduce the correspondence between the geometric problem and an algebraic problem, and a similar tactic will be used to analyze the surface area problem.

The other key problem introduces geometric representation, with each point inside triangle MAL corresponding to a number, the total distance to the three vertices. A short-term goal is to discover the property of the minimizing point. A long-term goal is for participants to use this tactic on other problems. For example, all rectangles with perimeter 24 can correspond to the points on the line with equation $x + y = 12$ where x and y are the length and width of the rectangle, and all rectangles with area 36 correspond to the points on the curve whose equation is $xy = 36$:

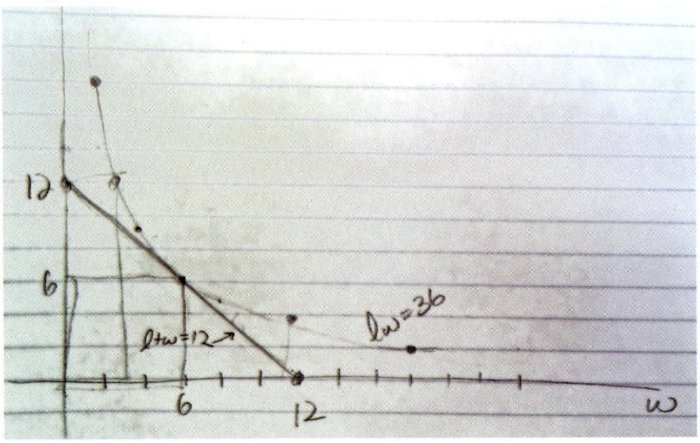

Problem Set 1 is also intended as an introduction to the style of the course. Consider having participants read the Introduction to learn about the course expectations.

Notes on Selected Problems

The opener is delayed in order to give participants opportunity to get geometry software working. If all participants have working geometry software, it is reasonable to have them work on the opener (at the end of the Important Stuff section) before working on the other problems, but it is not necessary.

In problem 1, try to press participants to give a better answer than "two variables and two equations". Give an example such as $x + y = 10, 3x + 3y = 30$ where this is not enough information, or an example such as $x + y = 10, x^2 + y^2 = 1$ where there is no solution.

Watch for participants making connections between problems 2 and 3, especially on part **e** where the solution is not an integer. Encourage participants to solve the problems without using the quadratic formula, especially if some participants do not have much experience with it.

In problem 5, watch carefully for participants who answer "no" by saying "three variables and two equations". This would only apply if the equations were linear, and these are not. Give an example such as $x^2 + y^2 + z^2 = 0$ where an equation can have a unique solution with more than one variable. Overall, encourage participants to spend a lot of time on this problem, even if you think they are not making much progress. One potential method is to pick a value for one of the dimensions, then proceed. It's likely this problem will remain unresolved in this set, and that is a good thing.

In the airport problem, show participants how to draw line segments, measure lengths, and calculate values based on measurements. When showing how to measure lengths, point out that other measurements are available (angle, slope, area, etc). The property of point X, that it forms three 120-degree angles from the segments, is verified by measuring the angles; this can also be used to better place the point. Some participants will look for the usual "centers" of a triangle:

- centroid: intersection of the medians
- orthocenter: intersection of the altitudes
- incenter: intersection of the angle bisectors
- circumcenter: intersection of the perpendicular bisectors

All of these centers are wrong. Participants should not be discouraged to find these centers, but should be encouraged to first find the "best" placement of point X for a specific triangle. It is likely that someone with no knowledge of these centers will take less time to find X and its property than someone with a lot of knowledge of the centers.

Problem Set 2

Goals of the Problem Set

Problem Set 2 cements the connection between rectangles and quadratics, offering a means of solving any quadratic equation by using "difference of squares". By knowing the maximum product for a given sum, you can calculate how far away the answers should be compared to the perfect pair.

Some problems in this set help participants toward understanding and recognizing complex numbers, both algebraically (problems 2d, 3h) and geometrically (problem 6). These are still previews and will be revisited when necessary, but look for participants making these connections early.

Notes on Selected Problems

For the opener, participants will need to learn how to place a point "on" an object (X on NL) and it may be useful to learn how to create a line instead of a segment. As before, measurement determines the property of point X (it forms equal angles at AXN and TXL). Look for a large number of potential explanations:

- By reflecting point T across NL to T′, the length AX+XT is minimized when AX+XT′ is minimized, but that can be a straight line. This is probably the "best" way to explain it, and can be hinted as by talking about an equivalent problem where T is on the other side of the river. Then, properties of vertical angles and congruent triangles prove the desired result.
- The total distance can be written as the sum of two square root functions and modeled on a graphing calculator or within the geometry software. As with the second solution, this is not a proof but a means of finding the correct placement of X.

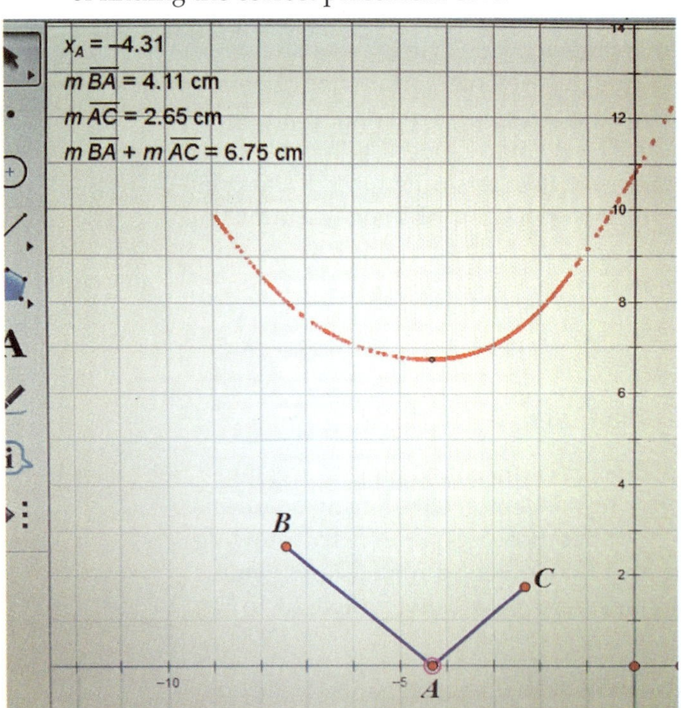

- By dropping perpendiculars from A and T, point X is "best" when the two triangles formed are similar. This is not a proof but can explain or find the correct X for a new situation.
- After dropping perpendiculars, connect A to the foot of the altitude from T to NL, and connect T to the foot of the altitude from A. The intersection point of these segments is directly above the correct placement of point X. In this case it can be proven that the two triangles formed are similar, then use this information to prove that the other two triangles (that include X) must also be similar.

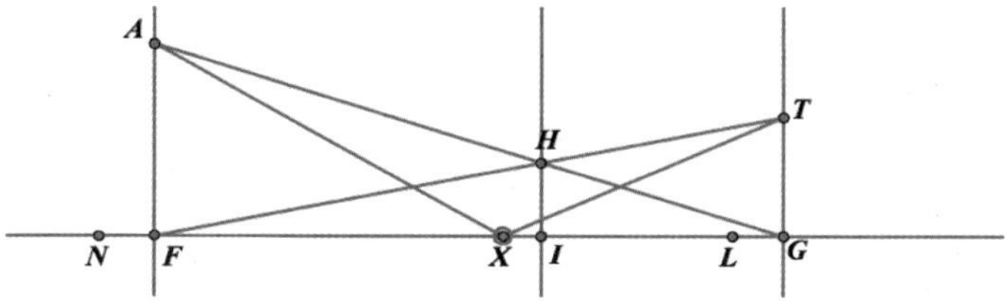

Carefully watch participants work on problem 2b. Those who struggle should be directed to problem 1, which gives the answer. Participants may not be used to working this way or actively looking for connections between problems.

Look for the methods participants are using in problems 3f–3h. As the sidenote says, discourage use of the quadratic formula even for those who know it well, and especially discourage teaching the quadratic formula from one participant to another. Some participants have used this problem as a jumping-off point to derive the quadratic formula, or at least a version of it for monic quadratics (with no leading coefficient).

Problem 4 is a quick check to see whether participants know the method of difference of squares, and a similar problem is found in Problem Set 3.

Problem 5 is important, since it encourages participants to think about dimensions other than integers. The 4-by-4 and 3-by-6 rectangles are easily found but the third is tougher, and many participants will discover a useful method for future work: pick a value for one of the dimensions then use that value to solve the rest of the problem. This is one means of attacking the surface area and volume problem when it returns.

Problems 6 and 7 are a preview of work with complex numbers. Ideally, participants should also work on problem 9, so they can see what the repeated transformation looks like. In particular, the application of the transformation twice will eventually be linked to the identity $i^2 = -1$ and the application of the transformation four times will be linked to $i^4 = 1$. It's unlikely these connections will be noticed here, though, so just watch for recognition of the 90-degree counterclockwise transformation.

Problem Set 3

Goals of the Problem Set

This problem set continues to focus on parameterization. Geometrically, parameterization can lead to the recognition of invariants, as it does in this set's opener, or the recognition of optimal and extreme cases, as in the previous sets' openers. Coordinates can be used in several ways, including verification of geometric properties (problem 3), visualization of dimensions in new ways (problem 4), and the building of functions to model situations (problem 6).

These concepts will be brought to bear on the surface area problem in the long term, but in the short term the concept can be used to find more rectangles like the ones asked about in Problem Set 2 whose perimeter and area are the same numeric value.

Notes on Selected Problems

The opener will take a long time to build. Lead participants through the construction of an equilateral triangle by building a line segment and two circles; from segment AN the first circle has center A and radius point N, and the second has center N and radius point A. Be especially careful that participants use the points when building these circles instead of "eyeballing" it: ideally, make the error yourself so that others can watch out for it.

This is also the best time to teach the concept of "hiding", which will be necessary both for hiding the circles and the perpendicular lines from point X. Show participants how to build one of the perpendicular segments by building the perpendicular line, marking the intersection with the base, hiding the line, then drawing the segment from X to the marked point. Participants should be able to finish the diagram.

Encourage participants to show all measurements in the diagram so that when the sum of the three lengths is found to be constant, it is not judged to be "broken".

Participants will want to prove what they've found! One useful observation is that when X is crammed into a corner, the total length looks just like an altitude of the triangle; this then establishes the value of the sum and can help discover one of several good proofs. Problem 1 from Problem Set 4 offers one proof, so let participants know they will get a second shot at completing this. Another good proof is found by building three equilateral triangles from point X that use the three segments as altitudes, then moving those triangles so they sit atop one another.)

Problem 2c is another that targets complex numbers. Encourage participants to say more here than that the situation is impossible. Their result in problem 2b can also be used.

In problem 4, encourage participants to test each point rather than try to write an equation for (x, y) immediately. The work with specific points should generalize to the result in problem 3d. Problems like this will continue to be asked with more and more difficult relationships, and by testing points, an equation can be constructed. This is a tactic that can be used effectively by participants when a formula is not available, or even when the goal is to build a formula. Several simpler situations, such as the collinearity of points on a line, can also be solved the same way.

Problem 4 is key as it helps participants explore parameter spaces. The first parameter space is simple: (l, w) are the dimensions of a rectangle. A more useful parameter space of length and area-to-perimeter ratio, (l, R), will be used in the next problem set. Participants who succeed on problem 4 may be able to use the technique to solve problems 7 and 9.

Problem 6 is a preview of a future piece on the method of calculating the line of best fit for data. In this case the "best fit" is a one-dimensional answer that gives the minimum total distance. This problem will be revisited for the square of distances, then again in two dimensions where the dimensions are the parameters of the line of best fit. Participants are likely to be surprised by the result, since most will expect Darryl's best spot to be closer to the mean of 1, 3, and 17. As it turns out, the best spot will always be the median of the data values.

Problem 8 is a re-expression of problem 6 from Problem Set 1, but it is still unlikely that participants will resolve this problem one way or the other. If any participants do find two different boxes with the same surface area and volume, consider hiding the information for a while to let others work on the problem in future sets.

Problem Set 4

Goals of the Problem Set

One major goal of this problem set is to revisit some of the previous geometric work, finding a way to construct the 120-degree point found in Problem Set 1 and potentially prove it using the diagram from Problem Set 3. It is also a chance for participants to solidify their skills on construction using geometry software.

A second major goal is to learn a technique for scaling geometric objects to make two calculations equal: the area and perimeter of rectangles, or the surface area and volume of boxes. Some participants may already have discovered the connection between these dimensions and reciprocals, but here the result is given in order for participants to discover why the connection exists.

Notes on Selected Problems

Problem 1 should be pretty straightforward; if anyone needs a push encourage them to draw a triangle with point X as a vertex. If some participants haven't yet learned the result from Problem Set 3 you might lead off with a participant's diagram and explanation.

If participants are having trouble building a right triangle, ask them where a right angle would come from. Remind participants about "hiding" and they should be fine with all parts. Measuring the area of the triangles should not be a problem, but some geometry software requires the construction of a "triangle interior" before measuring the area.

Don't worry if participants can't prove that the 120-degree point is minimal. This problem can be solved in multiple ways and it is fun to pursue, so even if someone does prove it here, keep the proof under wraps until later.

The result from problem 2 should be surprising but it should be resolvable pretty quickly: if $2W + 2L = LW$, then $\frac{1}{L} + \frac{1}{W} = \frac{1}{2}$ by dividing through. One interesting fact here is this quickly resolves the problem from Problem Set 2 about finding other rectangles with this property. An engineer might point out that L and W are the resistances of two resistors in parallel with a net resistance of 2 ohms.

Problem 3 is another parameter space, but a more interesting one. It isn't clear from the description that the graph of the parameter space should be linear, but it can be visualized then proven. The consequence of this is there must be *exactly one* value of l that gives $R = 1$ and that starting from any rectangle one could find a similar rectangle with any desired ratio.

Problem 4 then follows up on this idea for rectangular boxes. While not explicitly stated, some participants will follow the same method as in problem 3 to scale up or down, and once again there is exactly one valid box similar to the given box. And once again, the fractions will add up to $\frac{1}{2}$. In a future set participants will be asked to find all boxes with integer side lengths that have the equal surface area-volume property, and the relationship to unit fractions is key to successfully analyzing that problem. As with the perimeter-area problem, encourage the algebra that leads to the relationship: if $2IM + 2TM + 2TI = TIM$, then $\frac{1}{T} + \frac{1}{I} + \frac{1}{M} = \frac{1}{2}$ by dividing each part of the equation by $2TIM$.

Problem Set 5

Goals of the Problem Set

The primary goal of this problem set is to introduce complex numbers, giving participants the opportunity to see why a visual representation of complex numbers might be useful. Several problems ask participants to answer equivalent questions about complex numbers and about the coordinate plane, with the objective that participants may notice the connection and recognize the utility of graphing in the complex plane. It is important here that the participants build this representation instead of being told directly about it; ideally look for a participant with little or no experience with complex numbers who makes the connection. Watch out for participants trying to directly teach anything other than the basic operations about complex numbers, since the hope is for as many participants as possible to come upon the visual representation on their own.

Notes on Selected Problems

Problem 1 and the boxed geometry problem have a similar flavor. Participants should discover through their sketch that the product $AP \cdot AQ$ is constant regardless of the location of Q. Then, their observations in problem 1 can come into play: the sum is smallest when the two numbers are equal, and the sum is largest when one number is made as large as possible. Therefore, the maximum $AP + AQ$ occurs when PQ is a diameter, and the minimum occurs when P and Q are the same point. Some participants may be curious how to prove that $AP \cdot AQ$ is constant while others may know the theorem (called the "Power of a Point"). The proof uses similar triangles and the property of inscribed angles: draw two different possible placements for the line through A, P, and Q, then draw the chords connecting each P to the opposite Q. Inscribed angles and the shared angle at A force similar triangles, then the similar triangles have proportional sides. This proportion, multiplied out, shows that $AP \cdot AQ$ must be constant for any location of P or Q. Some later problems (problem 20 in this set, among others) explore the value of $AP \cdot AQ$ as a function of the location (coordinates) of point A, and there is a strong connection to the standard form of the equation of a circle.

Problems 2 and 3 are intended to cement the big ideas from Problem Set 4, so watch for participants who are still unclear on what to do.

Problems 5 and 6 use the same numbers: point S is $(2, 1)$ and $s = 2 + i$, while point M is $(4, 6)$ and $m = 4+6i$. Of particular interest are problems 6e and 6f, where multiplying by i give the same points that came from the rotation in problem 5. If anyone notices this, play it up, but don't worry if it doesn't happen.

Problems 7 and 8 use the same numbers, and are intended to eventually define the magnitude of a complex number.

In Problem 10, watch for some participants working out the entire expression algebraically to get $a^2 + b^2 = 25$. If this happens, terrific, and the connections to problem 7b and problem 9 can be established immediately. There will be more opportunities, though.

Problem Set 6

Goals of the Problem Set

A primary goal of this problem set is for participants to find and understand the utility of representing complex numbers graphically. Some will already have discovered this in Problem Set 5, and Problem Set 6 is then an opportunity to solidify and extend these ideas.

A secondary goal is for participants to learn how to calculate the height and area of a triangle given its three side lengths. Pointing out that SSS is a triangle congruency may be helpful here: since there is only one triangle with any three given side lengths, it makes sense that there should be a way to determine that triangle's area. While some participants may know a formula for this, we hope that by calculating the area with specific numeric examples, they will be able to *derive* this formula.

A final goal is for participants to work with Pythagorean triples. By Problem Set 9, participants will be able to construct Pythagorean triples using complex numbers, and use Pythagorean triples to construct triangles like the ones in problems 2 and 4 with integer sides and integer area.

Notes on Selected Problems

The opener should be pretty quick. Watch out that some participants may miss $(7, 4)$ while others may miss some of the switches and reflections. Later in the set, participants will work with $8+i$ and $7+4i$ as complex numbers, hopefully making the connection to the opener.

Watch out for participants missing the connection between problems 1 and 2. In particular, many participants may try to use a formula or some trigonometry in problem2, and that should be discouraged immediately. All the formulas participants might use here—the Law of Cosines and Heron's Formula in particular—can be derived from the work in these problem sets. Make liberal use of the sidenote here: "don't use anything with the letter s". That includes Heron's Formula and anything to do with sine or cosine.

Problems 5e and 7 are meant to dovetail, as are problems 7d and 8. While participants work on problem 8, look for them to reconnect the information with the opener. If you need to, remind them that a and b could be negative.

Many participants may miss the connection between problem 9b and 9c. You could ask them to repeat the calculation with a complex number like $(2+i)^2$ or $(3+2i)^2$, which lead to the more familiar Pythagorean triples 3-4-5 and 5-12-13.

Problems 11 and 12 are related: one of the solutions in problem 11 is 15, 6, 4 (the other is 12, 10, 3). The volume and surface area of the box with dimensions $\frac{1}{15}, \frac{1}{6}, \frac{1}{4}$ will both be equal to the volume and surface area of the box with dimensions $\frac{1}{12}, \frac{1}{10}, \frac{1}{3}$.

There is a lot of Important Stuff in this problem set, but most of it is quick work. Be mindful that some participants may not get through it all; it's alright if participants don't see problems 11 and 12, but if they don't make it past problem 8 they are likely to miss significant connections that are coming.

Problem Set 7

Goals of the Problem Set

The main goal of this problem set is an understanding of plotting numbers in the complex plane and some interesting results:

- What happens when you multiply a complex number by i? by i^2? by i^n?
- What happens when you square a complex number, especially to its magnitude?
- What patterns form when you take consecutive powers of a complex number?

A secondary goal is a connection of the earlier work from Problem Sets 3 and 4 to the area of a general triangle given its side lengths. The radius of the circle that sits inside a triangle, tangent to all sides, controls the ratio of the perimeter to the area. The relationship is $A = \frac{1}{2}rP$, and if $r = 2$ the triangle will have equal perimeter and area.

Notes on Selected Problems

The initial investigation determines the *incenter*, a point that is equidistant from all three sides of the triangle. By pouring a huge amount of salt on the triangle, it all falls off the closest edge, so the apex of the salt is the one point that is the same distance to all sides. The ridges are the angle bisectors, which some participants may remember as the set of points equidistant to two given lines.

Logistically, fine-grained salt is best, and triangles should come from heavy cardboard or plastic in order to support all the salt. Lay the triangle on something sturdy and tall, such as a coffee can or empty salt container, and lay the whole thing inside a large box or box top so that the excess salt does not make a mess.

If no materials are available, use the sketch provided. Try not to spend too much time on the initial investigation, but the hope is that participants can fill in the last question asked: the top of the salt heap is the same *distance* to the three *sides*. Participants may not realize that this means the point is also the incenter, yet.

The concept is developed further in problems 3 and 4. Problem 3 uses the triangle from problem 1, which has equal perimeter and area, asking participants to place the incircle. One option is to prepare a triangle of the right size (7-15-20, perhaps centimeters) then pour salt and verify that the center and distance are correct as determined in the software sketch. Problem 4 then asks participants to construct the argument that if the incircle radius is 2, the triangle's area and perimeter must be equal. Look also for any participants handling problem 3 in the same way they handled problem 1 from Problem Set 4, with colored labeling of the three triangles generated using the incenter as a vertex.

If you haven't yet, introduce the plotting of points in the complex plane before anyone works on problem 5. Try to make the introduction as short as humanly possible, and ideally have this done by a participant who has discovered the connection recently.

Problem 5 is important, focusing on the results when multiplying by i and its powers. Note that the same points here were used in Problem Set 2, and participants who did problem 9 in that set will have effectively already completed this problem geometrically.

Problems 6–10 all have the same general flavor. Problems 6 and 9 ask for plotting. Some participants may notice a "spiral" pattern to the powers of $1 + i$ and this is something to emphasize but difficult to see on the other examples here. If time permits, take another complex number with small magnitude, such as $1 + \frac{1}{2}i$, and look for the same spiral. The goals here are for participants to recognize:

- The magnitude of a power is the power of a magnitude: this is especially easy to spot in problem 8 but can also come from problems 6 and 7.
- The "direction" of a complex number seems to move consistently counter-clockwise from each power.

The second piece on direction is ill-formed at this point, because the importance of direction as a concept is not yet there. This will be followed up deeply in the next two problem sets.

In problem 9, look for any participant giving exact fractional values, since the results are very interesting: the numerators and denominators form Pythagorean triples! For example, z^2 gives the 7-24-25 triple, and z^3 gives the much-less-known 44-117-125 triple.

Problem Set 8

Goals of the Problem Set

The major goal of this problem set is to understand what happens when a complex number is squared, and use that result to build Pythagorean triples. This happened in some problems in Problem Set 7, and happens here in the opener along with problem 8.

A secondary goal is the meaning of the "direction" of a complex number. The opener is helpful, but problem 1 is especially good for showing the increasing direction of powers of a number, but also for coming up with a good definition of "direction". After working on problem 1, participants should get a feel for its definition as the angle with the positive real axis.

Notes on Selected Problems

The opener is pretty straightforward, but ask participants to continue generating examples as necessary, especially examples that push the square outside Quadrant I. This will happen for $a + bi$ with $a < b$. This is probably better than using negatives and having the original complex number outside Quadrant I, because a key relationship (the angle is doubled) is very hard to recognize when the original direction angle is large.

Problem 1 follows this same idea by using an initial z with a small direction angle. A common mistake has been made many times due to the round-off in this problem, leading many to state that $z^4 = i$ for this number. It turns out z^4 is very close to, but not exactly equal to, i. The actual z that would make $z^4 = i$ is a messy radical, while the actual value of z^4 here is $\frac{-239}{28561} + \frac{28560}{28561}$. Oh, and 239, 28560, 28561 is another Pythagorean triple!

Some participants may use trigonometry in problem 2, but let others know they can use geometry software to measure the angle; no trigonometry should be necessary for any problem in the entire course's Important Stuff.

Problem 4 is the first use of the word "direction", and problem 5 is the first use where something specific needs to be said about it. Make sure there has been a discussion around a good definition of direction before this; perhaps after everyone has worked on the opener or on problem 2.

Problem 8 is the key problem in the set. Ask participants to explain why it works in as much detail as possible, then let them loose to find some huge Pythagorean triples. Problem 20 gives the means to generalize, giving a formula that some participants may have seen or know of previously.

Problem Set 9

Goals of the Problem Set

The major goal of the problem set is the development of formulas for the area of a triangle given its three side lengths, and for the radius of the incenter of a triangle given its three side lengths. These classical formulas were learned by the Greeks using some difficult geometric properties, but participants can work through numeric examples to construct the formulas. The style of keeping track of steps in a numeric calculation, then generalizing, is a very valuable tool in all mathematics. The Common Core State Standards make this one of the eight Standards for Mathematical Practice, "Look for and express regularity in repeated reasoning."

A secondary goal is the construction of Heronian triangles, triangles with integer sides and integer area. Such triangles can be created by pasting together scaled versions of Pythagorean right triangles. While it is not proven here, any Heronian triangle is made of two scaled Pythagorean triangles in three different ways, since any of its three altitudes must have rational length, and the two pieces of the base must also have rational length.

Notes on Selected Problems

The Sketchpad file linked in the opener includes an additional menu of tools not present in a normal sketch. These tools are found on the last icon in the panel (it looks like a "play" button). It may take participants some time to get used to these tools, so they are introduced rather gently here. If you haven't yet, this is also an opportunity to use the "Locus" tool to show the path taken by v^2 as v changes. Show that option to anyone specifically looking for a visual path for v^2, or at the end of exploration.

In problem 2, watch for some participants dropping the altitude to the side of length 28, producing an immediate solution by dividing the triangle into an 8-15-17 and a 15-20-25 right triangle. While a nice idea, make sure these participants can also follow the steps to calculate the lengths and area, because the next example will be side lengths a, b, and c and there is no way to produce an immediate solution.

Let participants play with problem 1 as well as problems 7–9 a bit here. The techniques here can produce some really large-sided triangles that still work, and it's a fun time.

Participants with good algebra skills may want to attack problem 17, which asks them to manipulate the formula into the "semiperimeter" version that is usually given as Heron's Formula. The formula was written this way since Heron's proof involved constructed the length s, along with the others, into a single diagram, then using geometric means to construct another shape with the same area as the given triangle.

Problem Set 10

Goals of the Problem Set

There are several small goals for this problem set. One goal is a greater familiarity with solving equations through factoring, along with the rule for the vertex of a parabola. This information will be necessary in the study of the line of best fit, the major goal of the next three problem sets. Knowing the x-value of the vertex allows simple calculation of the value of parameters that produce the line of best fit, since the calculation of the sum of squared errors involves only quadratics.

Another goal is recognition of why squared errors might make sense. Problems 7 and 8 attack the same data using absolute error, then squared error, and the graphs of the sum of errors in each case. The graph when using absolute errors becomes more and more convoluted, while the graph when using squared errors remains a quadratic at all times. Additionally, the graph when using absolute errors will not have a single minimum when there are an even number of points used, while the graph when using squared errors always has a single minimum value.

A third goal is further familiarity with graphing in the complex plane. By focusing on a "Trace" or "Locus", the idea changes to a functional idea, with the magnitude circle as input and some other path as output. This idea will be further developed throughout to Problem Set 13, where the concept will be used to sketch a proof of the Fundamental Theorem of Algebra.

Notes on Selected Problems

The opener may prove challenging for some participants, especially as you try to teach how to make a trace or locus. Try to focus them on single values of ν, especially those which are easy to calculate such as $\nu = 2, \nu = 2i, \nu = -2, \nu = -2i$. Animating ν may also help participants understand the path the outputs take. Ideally spend the most time looking at ν^3, then $\nu^2 + \nu$ and $\nu^2 - \nu$. With ν^3, focus especially on the fact that the angle grows three times as fast, and "wraps around". Some participants may have solved cubic equations like $\nu^3 = 8$ earlier in the course, and can use the diagram to find those solutions.

In problems 2 and 3, note that the answers can be found by re-expressing the variable. For example, in problem 3b, replace 2x by M and the equation becomes $M^2 - 14M + 45 = 0$. Be mindful of participants attempting to teach unusual or specific methods for factoring non-monic quadratics; all parts of problem 3 can be solved through mindful substitution, and no special-purpose technique is needed.

Problems 4–6 attempt to help participants determine the sum and average of the roots of a quadratic, then show that $x = \frac{-b}{2a}$ gives the vertex of $f(x) = ax^2 + bx + c$. In the end, if some participants have trouble with this, it is alright to give them the formula, but you might try having a participant or two explain their reasoning or use numeric examples.

Problems 7 and 8 are very important to what comes next. Use a graphing calculator, website, or geometry software to define the function as it is built. In problem 7 the first function is $a(P) = |x - 3|$, then each is the sum of more absolute value functions. Watch out for participants who may think $|x - 3| + |x - 17| = |2x - 20|$, a numeric example ($x = 10$) may help them discover their error. The purpose here is to note that the sum of quadratic functions is a quadratic function, while the sum of absolute value functions is a mess.

Problems 9–11 offer some different ways to think about an ellipse. Encourage participants to use the same tactics they used in previous problem sets: test points to determine the general formula. If some participants are still jumping to the general formula, direct them to problem 12 where they are very unlikely to know the general formula.

Problem Set 11

Goals of the Problem Set

The goal of this problem set is to learn how to calculate a least-squares line of best fit. A specific data set is given; participants will determine the best possible lines of several different slopes. After finding that all such lines pass through the same point, the sum of squared errors for the line of slope m through this point can be determined as a function of m. This function is quadratic and has an absolute minimum, and that m gives the desired line.

A secondary goal is to compare this process to the process when absolute errors are used. While there may be a best fit line according to this method, it is not possible to compute a single best-fit line for many slopes, and there are data sets for which there are multiple "best fit" lines. (See problem 14 for an example.)

Notes on Selected Problems

Since participants will be working with this data repeatedly, hand out multiple copies of the data plot found on the last page of the set. The initial exploration is intended to be free-form, where participants can claim any reason that their line is "best". There are many possible results here.

At PCMI 2011, the instructors built a "badness" calculator that computed the sum of squared errors for each suggested line, reporting back the value of the badness (and not the method used to calculate it). This was mainly done for comedic value, but it also helped participants discover what causes the sum of squared errors to be high, and some participants were able to determine how the "badness" was being calculated through enough examples. This is also a good way to move from the opener to problem 1, where the sum of squared errors method is suggested and $y = x$ is the first tested line.

In problem 1, watch for participants making mistakes in calculating the sum of squared errors, such as subtracting the error on negative differences, failing to square the error, or adding errors before squaring. Problem 1c is important, since it leads to the observation that the sum of squared errors for any line with slope 1 is quadratic with b as the variable.

In problem 2, watch for participants making errors in calculations that lead to incorrect results. All the best-fit lines should pass through the point $(4, 5)$, the balance point of the data. Problem 3 is self-correcting as participants are likely to find a mistaken line, but if more than one mistake is made it is likely that the group will not see the shared intersection at $(4, 5)$.

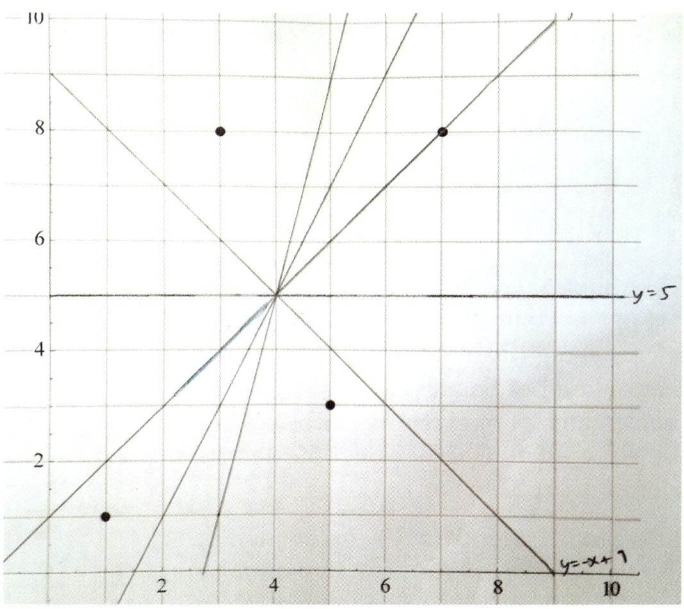

The data in problem 4 is the same as the original data, subtracting away $(4, 5)$. Don't tell this to participants; let them discover it, or reveal it only after working through problem 5.

Problem 6 can be ignored if you are short on time, but is a great way of demonstrating why the sum of squared errors is the standard method used. Like in Problem Set 10, the sum of the absolute value functions is a mess and has multiple "correct" lines. It turns out there *is* one best line for the entire data set, but the means of finding this line is somewhat different. The best line in terms of absolute error always passes through two of the data points.

Problem Set 12

Goals of the Problem Set

The primary goal of this problem set is to solve problems 3–5, finding all rectangular boxes with integer side lengths whose surface area and volume are equal in value.

There is no direct connection to the secondary goal of solidifying the work from Problem Set 11 on the line of best fit. Participants are asked to find the line of best fit for a given data set, and also asked to compute a large expression for the sum of squared errors given any line.

Another goal is the potential followup about parameterization that is possible by thinking about the variables m and b in two ways: the slope and intercept of a line in the xy-plane, and a point in the mb-plane. See the notes to problem 1.

Notes on Selected Problems

If participants have trouble starting the opener, ask them to pick a specific slope and calculate the best fitting line for that slope. After picking several slopes, participants will locate the balance point of data at $(5, 27)$ then determine the best possible value of m given this information. The most likely difficulty is failing to re-center the data as was done in problem 4 of Problem Set 11.

Ask participants to look carefully at the expression in problem 1b for any computable information. In particular the balance point $(5, 27)$ is visible in the first term: any line in the form $y = mx + b$ where $b + 5m - 27 = 0$ will pass through the point $(5, 27)$. The expression even tells you how much badness you add by being "off": if m is off by 1, the badness increases by 30, and then in proportion to the square of the error in m. If b is off by 1, the badness increases by 5, and then in proportion to the square of the error in b. At PCMI 2011 the instructors built an applet that would show the line of best fit but also all the lines of a specific badness larger than the best-fit badness. It's an interesting extension to explore this as a function of m and b, but also as graphed in the coordinate plane.

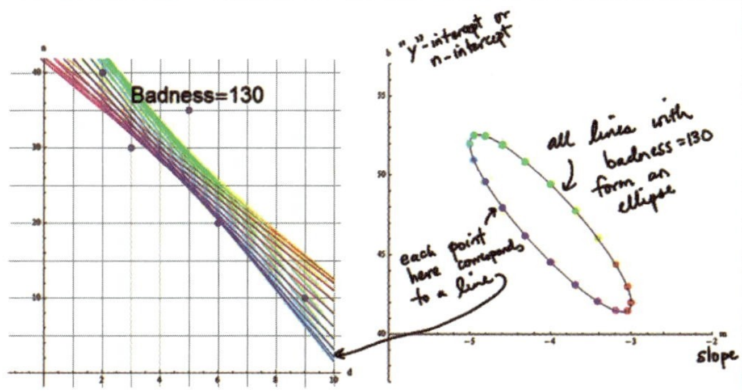

This kind of flexible and strategic use of algebraic and geometric representations gives us a way to prove that there is always a unique least squares line. Here's how:

- **The "badness" function is always a sum of squared linear functions, so m^2 and b^2 appear with a positive signs when the whole thing is expanded.**
- **Therefore, in the m-b plane, the curve of constant "badness" is an ellipse. The smaller the badness, the smaller the ellipse.**
- **Lowest badness = best fit line**
- **This happens when the ellipse collapse into one point.**

Problem 3 will take a long time, especially if you want participants to find all ten solutions. It is likely that participants will not take a systematic approach at first, but the sidenote offers a suggestion. The value of A must be between 3 and 6, and each choice of A limits the potential choices of B and C.

Participants are likely to recognize the equation in problem 4 as one they explored much earlier in the course, and problem 5 should be relatively simple based on the work in problem 3. Point out to participants that they were previously asked to solve problem 5, so the new representation of the equation as the sum of unit fractions was a breakthrough in finding solutions. Some participants may enjoy finding non-integer solutions.

Problem Set 13

Goals of the Problem Set

The primary goal of this problem set is to connect the work on boxes to a problem involving regular polygons meeting at a vertex. Participants will find the surprising connection then explore to discover the "why" behind it. A secondary goal in Neat Stuff is to continue exploring the graphs of polynomial functions in the complex plane, and sketch a proof of the Fundamental Theorem of Algebra.

Notes on Selected Problems

In the opener, participants are likely to find results by guess-and-check or by drawing pictures. The two simplest results are 3 hexagons (6-6-6), and a square with 2 octagons (4-8-8). There is probably no need to suggest any specific method of writing down the ways, since participants are likely to gravitate toward writing down the number of sides found. Eventually, participants will discover enough cases to recognize the data as the same data from problem 3 in Problem Set 12, and this should come as a pretty big surprise.

There are several ways to establish the connection. One way is to note that the angle measure of each interior angle in a regular n-gon is given by $180 - \frac{360}{n}$. If the three polygons have A, B, and C sides, the equation that must be solved is

$$180 - \frac{360}{A} + 180 - \frac{360}{B} + 180 - \frac{360}{C} = 360$$

This can be quickly simplified to

$$-360\left(\frac{1}{A} + \frac{1}{B} + \frac{1}{C}\right) = -180$$

The result follows. Note to participants that this is much easier to find when you know what you're looking for! A second version can be found using the formula for exterior angles, but this is less obvious. It ends up leading to the same result in one fewer step.

There are two problems numbered 1 in this set, and this is a deliberate typo, since the second "problem 2" is identical to the first. This time, the four interior angles must add to 360 degrees, leading ultimately to the equation

$$-360\left(\frac{1}{A} + \frac{1}{B} + \frac{1}{C} + \frac{1}{D}\right) = -360$$

Participants may end up solving the second "problem 2" while working on the first, since they may find it simpler to work with the unit fractions. The unit fractions make some choices obvious: for example, G must be 3 unless all variables are 4; geometrically, this means there are either 4 squares or at least 1 triangle.

Tilings that don't quite make it to 360 degrees, like the sample tiling with two hexagons and a square, still correspond to something interesting: they are possible tilings of a geometric solid. The sample creates a truncated octahedron, but there are many more possibilities, and the angle "shortage" can tell you the number of vertices of such a solid if it exists. See problem 14 for more about this. Tilings that overlap could potentially tile the hyperbolic plane (see problem 15).

Problem 4 revisits the initial geometric exploration from Problem Set 1, this time with four points instead of three. Remind participants if they ask that the best result in Problem Set 1 was a point that formed 120-degree angles when connected to all the given points. Here, that is not possible, but it offers some suggestions about what to do. A good solution is to place point X in the middle of the diagram, effectively drawing the two diagonals of the square. This connects all points more efficiently than connecting 3 of the 4 sides of the square. But the best solution adds *two* points that behave like the single point in Problem Set 1, built in a way so that all angles are 120 degrees. An excellent video illustrating this problem and its solution using soap bubbles is available at

`http://www.youtube.com/watch?v=dAyDi1aa40E`

Problem Set 14

Goals of the Problem Set

The goal of this problem set is to review the ideas of the entire course. This problem set is optional. You may prefer to have participants write their own review problems for this set.

Notes on Selected Problems

All problems are review problems, and they are roughly presented in the order they might have appeared in the rest of the course. The opener is a review of the Fundamental Theorem of Algebra and the method of proof discovered in Problem Set 13.

CHAPTER

3 Solutions

Problem Set 1

1. Yes. There are many possible ways to think about this. One is to look at the graphs of perimeter (a line) and area (a curve) in the coordinate plane of lengths and widths. These graphs can have at most two intersections, and those intersections correspond to the same rectangle.

   Another is to use the quadratic formula to solve the equations $P = 2l + 2w$ and $A = lw$ for either l or w, giving a rule that will determine the specific dimensions for given perimeter and area.

   Note that not all options are valid: for example, there is no rectangle with perimeter 10 and area 100.

2. **a.** 6 by 6
   b. 7 by 5
   c. 8 by 4
   d. 9 by 3
   e. This one is tougher. By using the pattern or by finding the solution to the system of equations, the dimensions are $6 + \sqrt{2}$ and $6 - \sqrt{2}$.
   f. There is no rectangle with these dimensions.

   Note: When the perimeter is n less than 36, the dimensions are $6 \pm \sqrt{n}$.

3. The answers here correspond to the answers in Problem 2.

   a. 6
   b. 7 and 5
   c. 8 and 4
   d. 9 and 3

- **e.** $6+\sqrt{2}$ and $6-\sqrt{2}$
- **f.** No real solution exists. Some participants may give the answers $6+i$ and $6-i$ but that is not required yet.

4. We can use the results from a previous problem here. Notice we found rectangles with perimeter 24, which was double 12, the coefficient of x. Thus we want numbers that add to $\frac{36}{2}=18$. The product of such numbers would then give us the area. Possibilities are:
$17 \times 1, A = 17$
$16 \times 2, A = 32$
$15 \times 3, A = 45$
$14 \times 4, A = 56$
$13 \times 5, A = 65$
$12 \times 6, A = 72$
$11 \times 7, A = 77$
$10 \times 8, A = 80$
$9 \times 9, A = 81$
In addition, rational and irrational values may be used, such as: 13.5×4.5, $A = 60.75$. Note that the area values will fall in the interval $0 < A \leqslant 81$.

5. If we know the surface area, S, and the volume, V, then we have the equations $S = 2(lw + lh + wh)$ and $V = lwh$. There are too many unknowns! If you knew one of the dimensions, then you could solve for the other two. In addition, there could be multiple answers. It is possible to find specific boxes with the property.

   Though you cannot necessarily determine the dimensions with confidence, you could find some boxes that meet Chance's given surface area and volume . . . so we will not make him sad!

   Don't worry, we will see this problem again soon!

Opener

Explore this problem! Try out everything that comes to mind. Utilize GSP's measuring capabilities - segments, angles, sum of distances, bisectors, etc.

Note that the lengths of the triangles sides will vary depending on your construction.

Interestingly the sum is smallest when each angle around vertex X right around 120°. Hmmm . . .

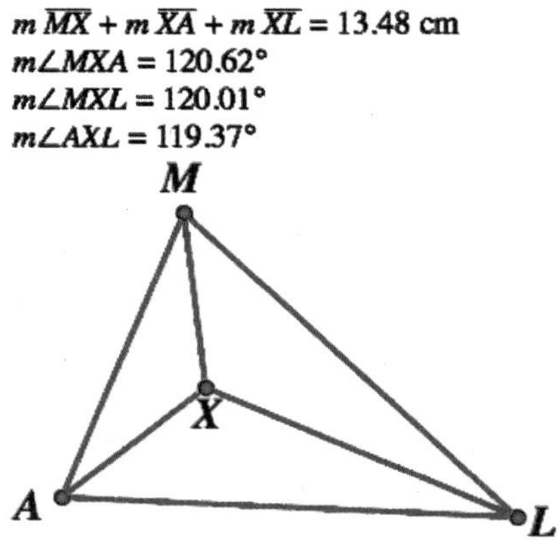

6. **a.** The distance from $(7,1)$ to $(7,10)$ is $\sqrt{81} = 9$
The distance from $(-2,9)$ to $(7,10)$ is $\sqrt{82}$, so the distance is not the same.

   b. The distance from $(7,1)$ to $(-1,1)$ is $\sqrt{64} = 8$
The distance from $(-2,9)$ to $(-1,1)$ is $\sqrt{65}$, so the distance is not the same.

   c. The distance from $(7,1)$ to $(-17,-17)$ is $\sqrt{900} = 30$
The distance from $(-2,9)$ to $(-1,1)$ is $\sqrt{901}$, so the distance is not the same.

   d. The distance from $(7,1)$ to $(-2,0)$ is $\sqrt{82}$
The distance from $(-2,9)$ to $(-2,0)$ is $\sqrt{81} = 9$, so the distance is not the same.

7. Notice that if a rectangle has the same value for both its perimeter P and area A, we have $2l+2w = lw \Rightarrow 2w - lw = -2l \Rightarrow w = \frac{-2l}{2-l} = \frac{2l}{l-2}$.
You can chose a value for l or w and then solve for the other. Make sure you choose values $l > 2$.
Also notice that $2l + 2w = lw \Rightarrow \frac{2(l+w)}{2lw} = \frac{lw}{2lw} \Rightarrow \frac{1}{w} + \frac{1}{l} = \frac{1}{2}$. Wow! That's Fancy Nice! The reciprocals of the dimensions sum to $\frac{1}{2}$. So . . . *Let's get ready to rectangle!*
4×4. $A = P = 16$
3×6. $A = P = 18$
$8 \times \frac{8}{3}$ gives $A = P = \frac{64}{3}$
$10 \times \frac{5}{2}$ gives $A = P = 25$
$7 \times \frac{14}{5}$ gives $A = P = \frac{98}{5}$
$18 \times \frac{9}{4}$ gives $A = P = \frac{81}{2}$
$20 \times \frac{20}{9}$ gives $A = P = \frac{400}{9}$
and so on . . .

8. This is similar to the previous problem, but we have another dimension. Some solutions can be found by picking values for two of the dimensions then solving for the remaining dimension. Some may notice that the equation $lwh = 2(lw + lh + wh)$ can be rewritten as $\frac{1}{2} = \frac{1}{h} + \frac{1}{w} + \frac{1}{l}$.

   Some specific boxes that work are 6-by-6-by-6, 8-by-8-by-4, and 12-by-12-by-3.

9. No it is not. There could be more than one triangle meeting Rina's criteria. For example, if Rina said she had a triangle with area $A = 210$ and perimeter $P = 70$, triangles with sides 17, 25, 28 and 29, 20, 21 both will have said area and perimeter.

Problem Set 2

Opener

In GSP you can create this picture in a way so that you can move the point X around, which is way cool!

Hopefully you notice that as you move X from the left to the right you will see the distance decrease and then begin to increase again. The key is to find the *sweet spot* for X. Let's not bring in calculus or other heavy powered tools.

Isn't the shortest distance between any two points a straight line? Hmmm . . .

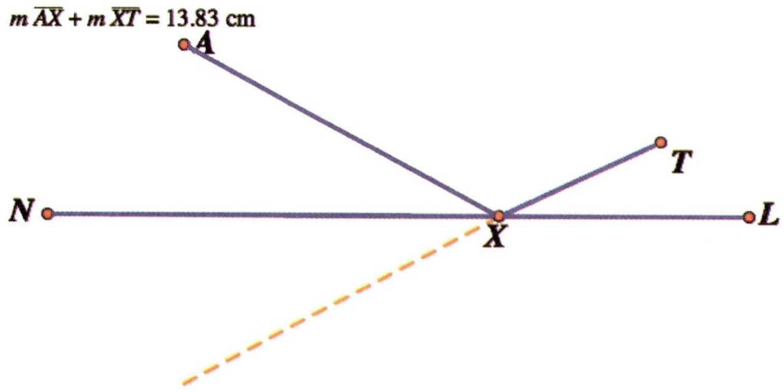

1. a. $2(11) = 22$
   b. $11^2 - (\sqrt{2})^2 = 121 - 2 = 119$
   c. $11^2 - x^2$

2. Each part of this problem is like a problem from Set 1. We want 2 numbers that sum to $22 = \frac{1}{2}(44)$ and multiply to be 117, 119, 100, 130. Use the square 11×11 to start, since $11 + 11 = 22$ and $11 \times 11 = 121$.

  a. $121-117 = 4$, so the dimensions are $11 \pm \sqrt{4} = 11 \pm 2 = 13, 9$
  b. $121 - 119 = 2$, so the dimensions are $11 \pm \sqrt{2}$
  c. $121-100 = 21$, so the dimensions are $11 \pm \sqrt{21}$
  d. $121 - 130 = -9$, so the dimensions are $11 \pm \sqrt{-9} = 11 \pm 3i$. Nonreal answer, so no such rectangle exists.

3. We can tie this problem to the previous problem. The previous problem wanted rectangles with perimeter $P = 2(l + w) = 44 \Rightarrow l + w = 22$ and area $A = 117, 119, 100, 130$. Here we are solving for the same kinds of numbers. We want numbers that sum to 22 and multiply to be 121, 120, 117, etc.

  a. $w^2 - 22w + 121 = (w - 11)^2 = 0 \Rightarrow w = 11$
  b. $121-120 = 1$, so $w = 11 \pm \sqrt{1} = 11 \pm 1 = 12, 10$
  c. $121-117 = 4$, so $w = 11 \pm \sqrt{4} = 11 \pm 2 = 13, 9$
  d. $121 - 119 = 2$, so $w = 11 \pm \sqrt{2}$
  e. $121 - 100 = 21$, so $w = 11 \pm \sqrt{21}$
  f. $121 - 2 = 119$, so $w = 11 \pm \sqrt{119}$
  g. $121 - (-408) = 529$, so $w = 11 \pm \sqrt{529} = 11 \pm 23 = 34, -12$
  h. $121 - 130 = -9$, so $w = 11 \pm \sqrt{-9} = 11 \pm 3i$

4. To sum to 200, you can write your two numbers in the form $(100 + x)$ and $(100 - x)$. Their product is $(100 + x)(100 - x) = 9991 \Rightarrow 9 = x^2 \Rightarrow \pm 3 = x$
So our two numbers are 103 and 97.
We can also relate this to the previous problem, if we want . . . You can view this as trying to solve the polynomial $x^2 - 200x + 9991 = 0$.
Let's use previous ideas! $x^2 - 200x + 10000 = (x - 100)^2$ $10000 - 9991 = 9$, so the solution to $x^2 - 200x + 9991 = 0$ is $x = 100 \pm \sqrt{9} = 100 \pm 3 = 103, 97$.

5. So we want three rectangles such that: $P = A \Rightarrow 2(l + w) = lw \Rightarrow l = \frac{2w}{w-2}$, $w > 2$. A value for w can be chosen, and then you can find the value of l. Also notice, though, that $2(l + w) = lw \Rightarrow \frac{1}{2} = \frac{1}{w} + \frac{1}{l}$. So you can find two numbers such that their reciprocals sum to $\frac{1}{2}$. Some possible rectangles include:

   4×4 gives $A = P = 16$
   3×6 gives $A = P = 18$
   $8 \times \frac{8}{3}$ gives $A = P = \frac{64}{3}$
   $10 \times \frac{5}{2}$ gives $A = P = 25$
   $7 \times \frac{14}{5}$ gives $A = P = \frac{98}{5}$
   $18 \times \frac{9}{4}$ gives $A = P = \frac{81}{2}$
   $20 \times \frac{20}{9}$ gives $A = P = \frac{400}{9}$
   and so on . . .

6. **a.** The graph of $\triangle SAM$ is below:

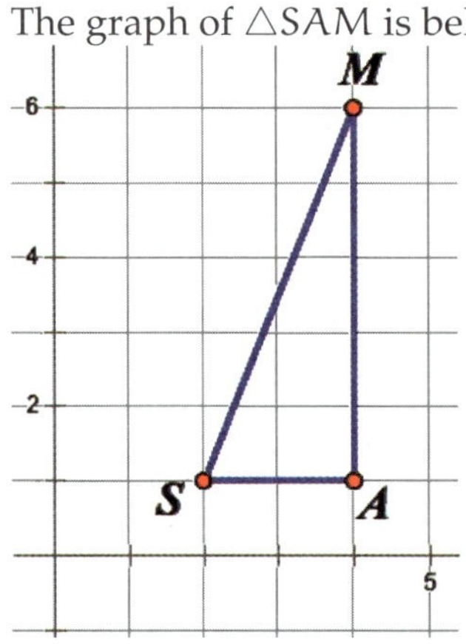

   b. $\triangle SAM$ was rotated about the origin by 90° to get $\triangle S'A'M'$ pictured below:

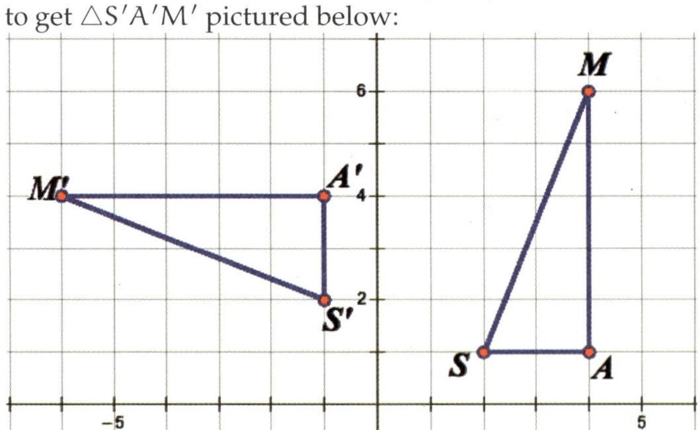

7. You can see from the graph below that $\triangle SAM$ was reflected about the origin.

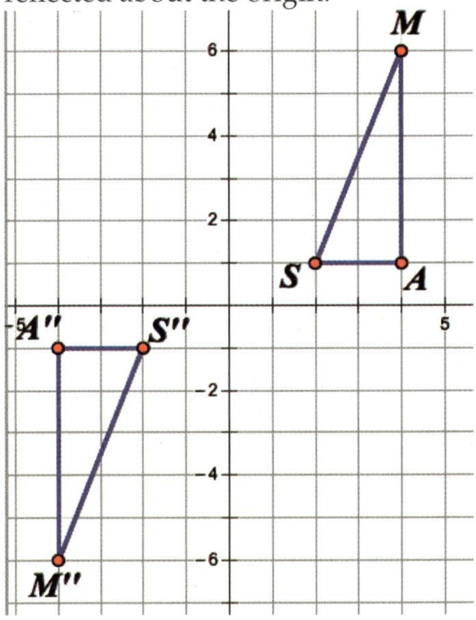

8. **a.** The distance from $(14,-4)$ to $(-2,-5)$ is $\sqrt{257}$, and the distance from $(14,-4)$ to $(4,11)$ is $\sqrt{325}$.

The distances are *not* the same.

b. The distance from $(-10,7)$ to $(-2,-5)$ is $\sqrt{208}$, and the distance from $(-10,7)$ to $(4,11)$ is $\sqrt{212}$.

The distances are *not* the same.

c. The distance from some point (x,y) to $(-2,-5)$ is $\sqrt{(x+2)^2+(y+5)^2}$ and the distance from ome point (x,y) to $(4,11)$ is $\sqrt{(x-4)^2+(y-11)^2}$.

If these distances were to be the same, then we would want to know when $(x+2)^2+(y+5)^2=(x-4)^2+(y-11)^2$.

Hey! These are circles! The first one is a circle centered at $(-2,-5)$ and the second one is a circle centered at $(4,11)$. Let's expand everything out and see what happens!

$$x^2+4x+4+y^2+10y+25=x^2-8x+16+y^2-22y+121$$

$$12x+32y=108 \Rightarrow y=-\frac{3}{8}x+\frac{27}{8}$$

All points (x,y) on the line $y=-\frac{3}{8}x+\frac{27}{8}$, and *only* those points, will be equidistant from $(-2,-5)$ and $(4,11)$. Geometrically, this line is the perpendicular bisector of the segment between the two points.

9. **a.** Applying the transformation twice maps $(x, y) \mapsto (-x, -y)$, which will reflect $\triangle SAM$ about the origin.
   b. Applying the transformation three times maps $(x, y) \mapsto (y, -x)$, which will rotate $\triangle SAM$ $+270°$ or $-90°$ about the origin.
   c. Applying the transformation four times maps $(x, y) \mapsto (x, y)$, which keeps $\triangle SAM$ in place. It will not move. We call this an identity map!
   d. Applying the transformation five times will be the same as applying the transformation once, since the fourth time does not do anything to $\triangle SAM$.
   e. Applying the transformation thirteen times will be the same as applying the transformation once, since performing the transformation in multiples of 4 does not do anything to $\triangle SAM$.
   f. 101 divided by 4 is 25 with a remainder of 1. Thus applying the transformation 101 times will be the same as applying the transformation once.
   If you remember your modular arithmetic, you can say $101 \bmod (4) = 1$.

10. **a.** Some possible points are plotted below:

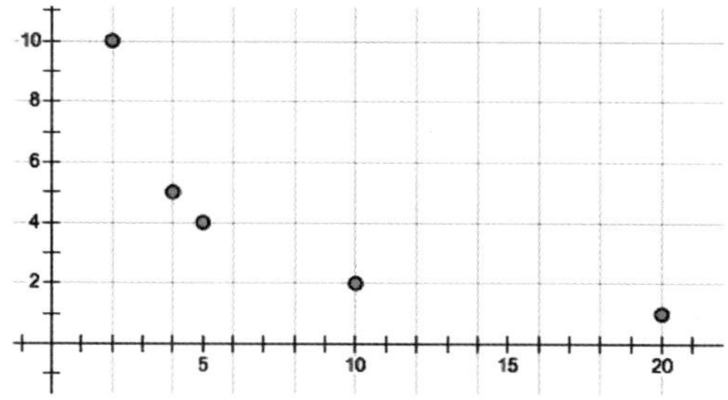

b. If our length and width can be any positive real numbers, then we can have infinitely many rectangles with area 20. $20 = lw \Rightarrow \frac{20}{l} = 2$. The graph of $\frac{20}{l}$ is below:

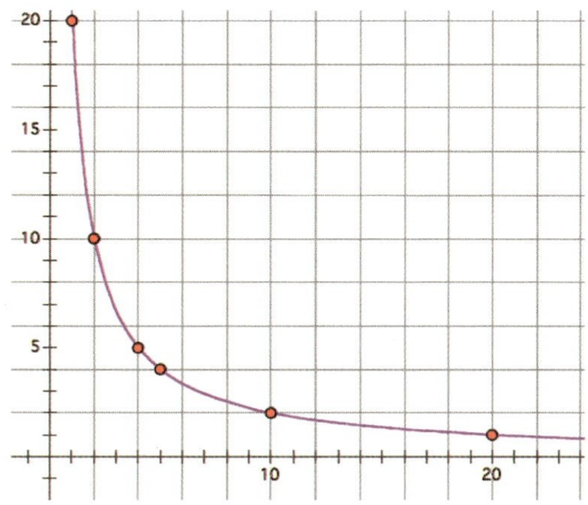

We can see that as values of l approach 0, values of w get larger so that the area is 20. Similarly, as the values of l get larger, the values of w get smaller and approach 0.

c. If our length and width can be any positive real numbers, then we can have infinitely many rectangles with perimeter 20. However, as you can see from the graph below, all the values must lie between 0 and 10 to ensure that the perimeter is 20.

The function in the graph is $w = 10 - l$. This is the result of working with the perimeter equation. $20 = 2l + 2w \Rightarrow 10 - l = w$.

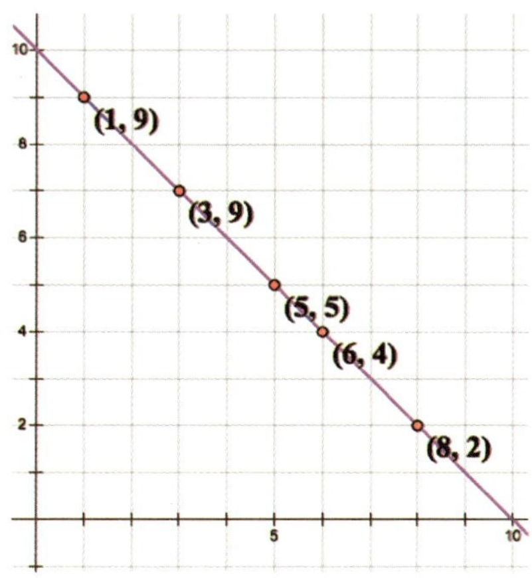

d. If we want a rectangle with both area and perimeter to be 20, then we want to know where our area and perimeter graphs intersect. In other words, when does $\frac{20}{l}$ equal $10 - l$?
$\frac{20}{l} = 10 - l \Rightarrow 20 = 10l - l^2 \Rightarrow l^2 - 10l + 20 = 0$
$l = \frac{10 \pm \sqrt{100-80}}{2} = \frac{10 \pm 2\sqrt{5}}{2} = 5 \pm \sqrt{5}$
The graphs illustrating the solutions are below:

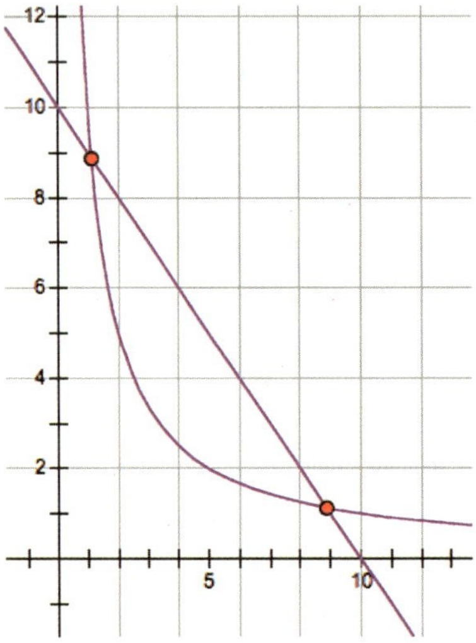

So there are two possible rectangles, one with either a length of $l = 5 + \sqrt{5}$ and width $w = 5 - \sqrt{5}$ or one with a length of $l = 5 - \sqrt{5}$ and width $w = 5 + \sqrt{5}$. Though, one could consider this to be just one rectangle.

11. In the situation where the area and perimeter both equal 12, the two graphs of $2l + 2w = 12$ and $lw = 12$ do not intersect. Therefore there is no rectangle with area 12 and perimeter 12.

12. **a.** Let P be at the point x on the number line. If Q is an arbitrary point on the number line, then the distance from some point P to Q is $f(P) = |x - Q|$ and the graph would look something like:

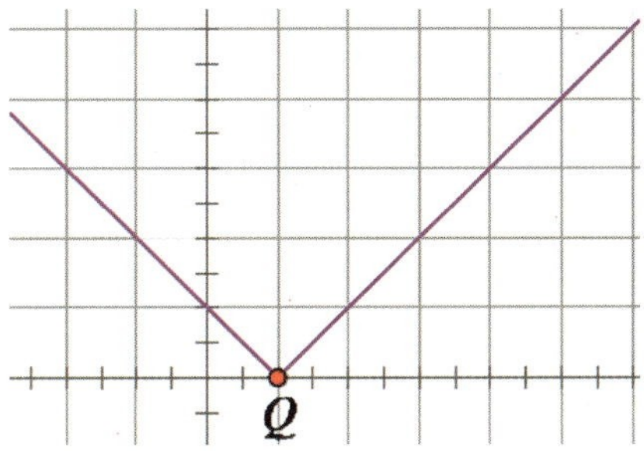

b. For total distance, we sum the distances from P to Q and from P to R and we get $f(P) = |x - Q| + |x - R|$ and the graph would look something like:

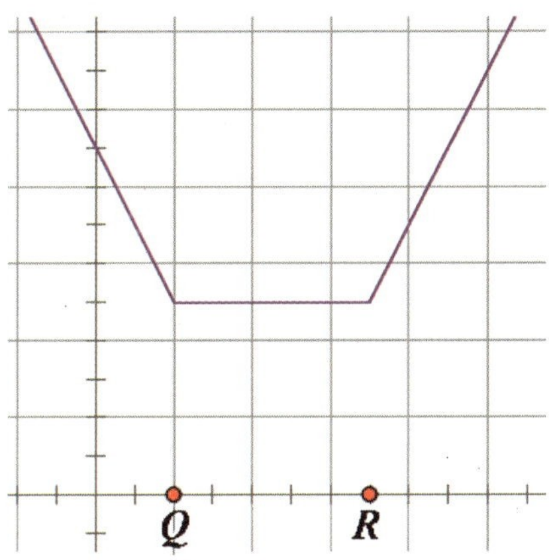

c. For total distance, we sum the distances from P to Q, from P to R, and from P to S and we get $f(P) = |x - Q| + |x - R| + |x - S|$ and the graph would look something like:

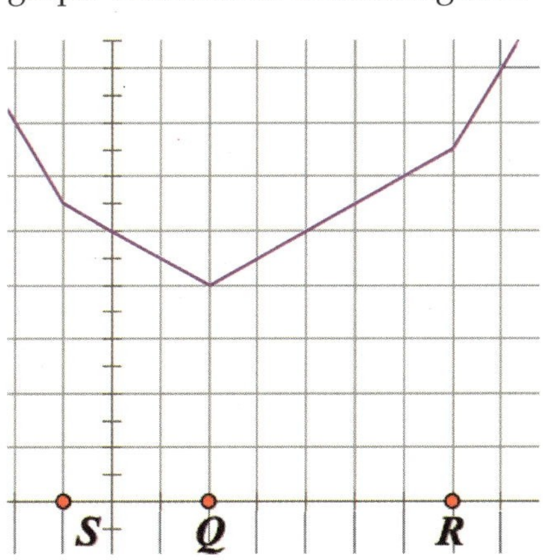

13. The triangle below has both an area and perimeter of 60.

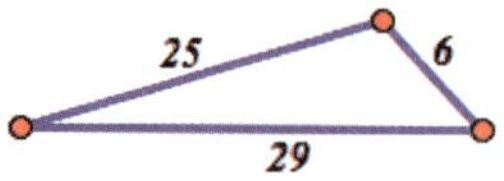

The triangle below has both an area and perimeter of 42.

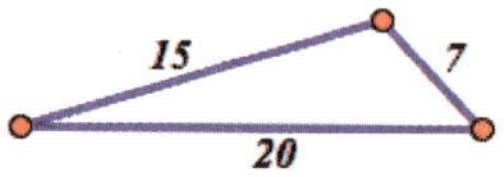

14.
- **a.** The remainder is -84
- **b.** The remainder is -76
- **c.** The remainder is -50
- **d.** The remainder is 0
- **e.** The remainder is 80

15. The remainders from the long division problem are the same as the values in our table! Whoa!

x	$f(x)$
0	-80
1	-84
2	-76
3	-50
4	0
5	80

16. Based on what we saw in problems 14 and 15, we can just evaluate $f(x)$ at values rather than using long division on a 12th degree polynomial.

- **a.** The remainder when dividing $f(x)$ by $x-1$ is equivalent to evaluating $f(1)$. $f(1) = 3$.
- **b.** The remainder when dividing $f(x)$ by $x-2$ is equivalent to evaluating $f(2)$. $f(2) = 4,101$.
- **c.** The remainder when dividing $f(x)$ by $x+1 = x-(-1)$ is equivalent to evaluating $f(-1)$. $f(-1) = -3$.

Problem Set 3

Opener

Okay . . . GSP can construct perpendicular lines, which is very nice! Moving X around now . . . WHAT?! The sum isn't changing! How could that be?!

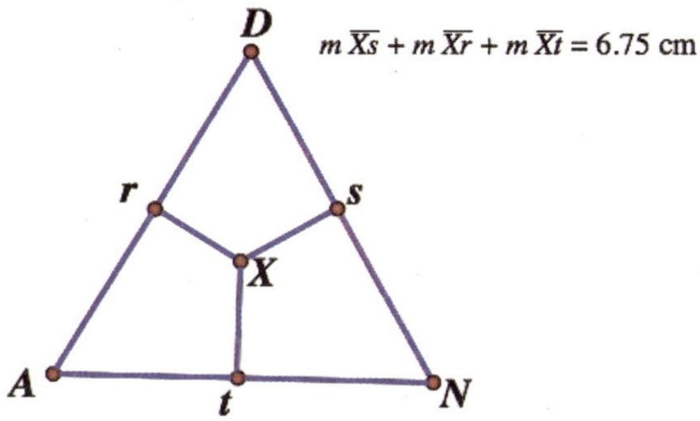

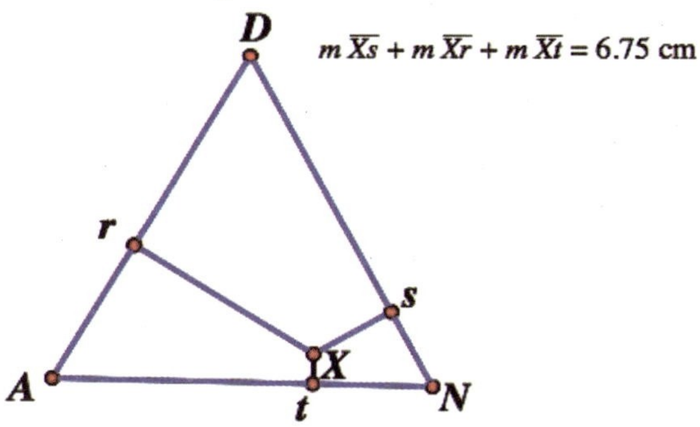

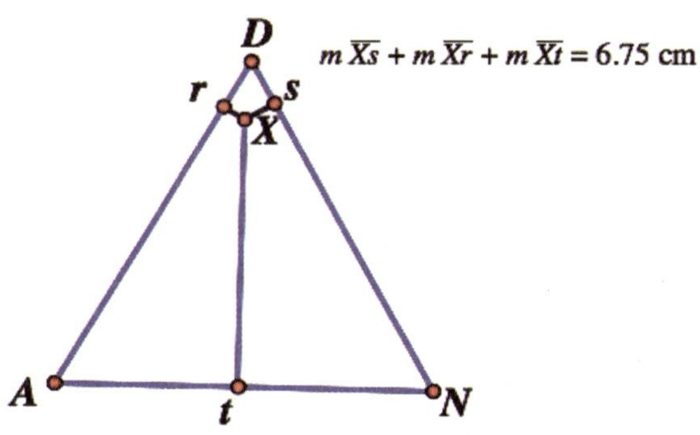

1. We can tie this problem to ideas from Problem Sets 1 and 2. To sum to 200, you can write your two numbers in the form $(100+x)$ and $(100-x)$. Their product is $(100+x)(100-x) = 9919 \Rightarrow 81 = x^2 \Rightarrow \pm 9 = x$. So the numbers are 109 and 91. You can also think about it as a quadratic: $x^2 - 200x + 10,000$, which has solutions $100 \pm \sqrt{0} = 100$. $10,000 - 9,919 = 81$, so the solutions are $100 \pm \sqrt{81} = 100 \pm 9 = 109, 91$. 109 and 91 sum to 200 and have product 9,919.

2. **a.** $15+15 = 30$ and $15 \cdot 15 = 625$. $625 - 161 = 64$, so the solutions are $15 \pm \sqrt{64} = 15 \pm 8 = 23, 7$. 23 and 7 sum to 30 and have product 161.
   b. $225 - (225 - n) = n$, so the numbers are $15 \pm \sqrt{n}$.
   c. $225 - 289 = -64$, so the numbers are $15 \pm \sqrt{-64} = 15 \pm 8i$.

3. **a.** No. The distance from $(13, -7)$ to $(5, 3)$ is $\sqrt{(5-13)^2 + (3+7)^2} = \sqrt{164} \neq 10$.
   b. Yes. The distance from $(13, -7)$ to $(7, 1)$ is $\sqrt{(7-13)^2 + (1+7)^2} = \sqrt{100} = 10$.
   c. No. The distance from $(13, -7)$ to $(10, -16)$ is $\sqrt{(10-13)^2 + (-16+7)^2} = \sqrt{90} \neq 10$.
   d. The distance from $(13, -7)$ to (x, y) is $\sqrt{(x-13)^2 + (y+7)^2}$. For (x, y) to be 10 units away from $(13, -7)$, we want all (x, y) such that

$$\sqrt{(x-13)^2 + (y+7)^2} = \sqrt{100}.$$

   Square both sides: $(x-13)^2 + (y+7)^2 = 100$ All points (x, y) on the circle with center $(13, -7)$ and radius 10 will be 10 units away from the point $(13, -7)$.

4. □RAUL corresponds to the point $(10, 15)$.
   a. If a rectangle is similar to □RAUL, then there is a common scale factor $s > 0$ applied to all sides of the rectangle. Thus any point of the form $(10s, 15s)$ will correspond to a rectangle that is similar to □RAUL. Some possible examples are $(2, 3)$, $(20, 30)$, $(8, 12)$, $(50, 75)$, $(\frac{1}{3}, \frac{1}{2})$, $(1, \frac{3}{2})$, $(\frac{2}{3}, 1)$, etc. See some more in the graph below!

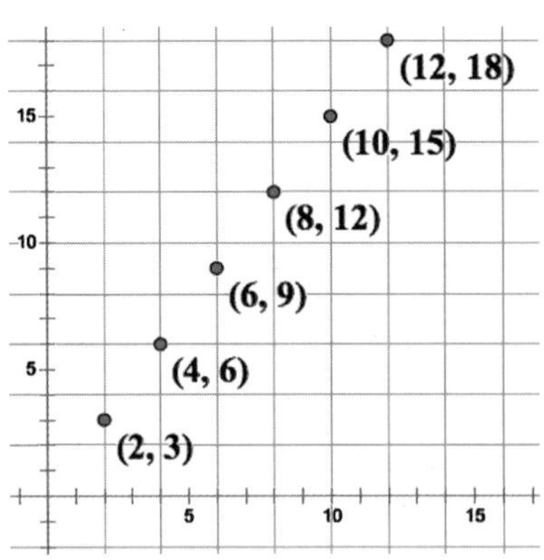

b. Notice that the slope between $(10, 15)$ and $(10s, 15s)$ is constant: $m = \frac{15s-15}{10s-10} = \frac{15}{10} = \frac{3}{2}$. Any point on the line $w = \frac{2}{3}l$, with $l, w > 0$, will correspond to a rectangle similar to $\square RAUL$.
Notice that all the points in the previous graph fall on this line!

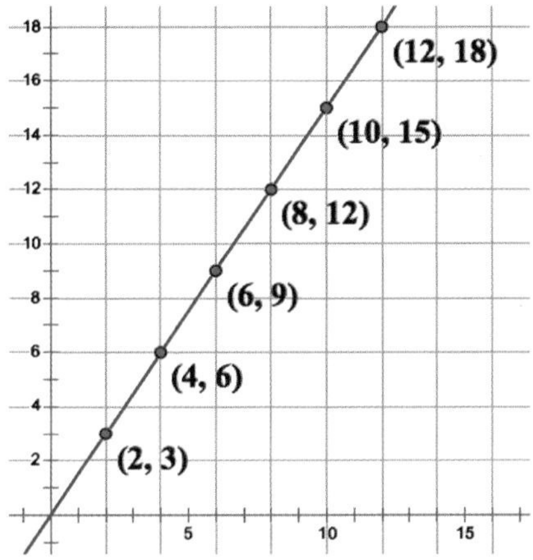

c. For a rectangle similar to $\square RAUL$, we have $l = 10s$ and $w = 15s$. In order for the perimeter to be larger than the area, we need to find the scale factor(s), s, that satisfy the equation:
$2(10s + 15s) > 10s \cdot 15s$
$50s > 150s^2$
$0 > 3s^2 - s$
$0 > s(3s - 1)$
Values of s in the interval $\left(0, \frac{1}{3}\right)$ will result in rectangles similar to $\square RAUL$ with perimeter greater than area. For example, $s = \frac{1}{5}$ gives $l = 2, w = 3$, which has $A = 6$ and $P = 10$.

d. We want area to equal perimeter, so we have:
$10s \cdot 15s = 2(10s + 15s) \Rightarrow 150s^2 = 50s \Rightarrow$
$3s^2 - s = 0$
$s = 0, \frac{1}{3}$
A scale factor of $\frac{1}{3}$ gives $l = \frac{10}{3}$ and $w = 5$. Then $A = P = \frac{50}{3}$.

5. Answers will vary. Typical answers might be the mean of the numbers; it is not typical that the correct answer of $(3, 0)$ will be guessed, which is the eventual point.

6. a. $f(P) = |x - 3|$ and the graph is below:

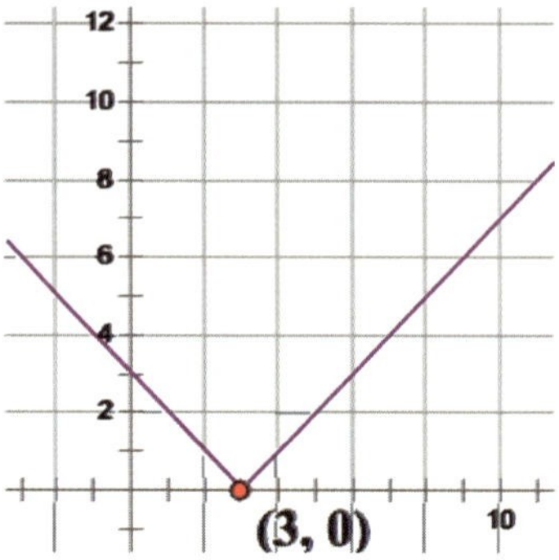

b. $f(P) = |x - 3| + |x - 17|$ and the graph is below:

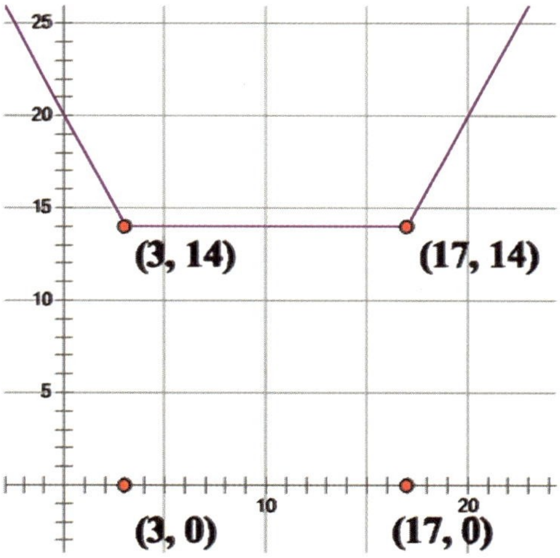

c. $f(P) = |x - 3| + |x - 17| + |x - 1|$ and the graph is below:

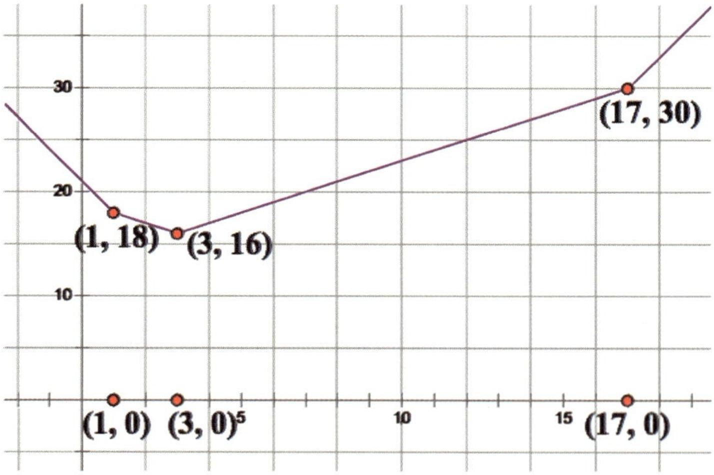

d. Looking at the graph in the last part, you can see that Darryl should live at the point $(3, 0)$ in order to travel the minimum distance.

7. A few possibilities are $6 \times 6 \times 6$, $4 \times 8 \times 8$, $3 \times 12 \times 12$, $3 \times 10 \times 15$, $5 \times 5 \times 10$, but there are many more.

8. This is a repeat of the problem from Set 1, and participants may or may not have found a valid pair of boxes yet or made a convincing argument that such a pair may exist. One simple pair of boxes is $4 \times 10 \times 15$ and $5 \times 6 \times 20$.

9. We did this in Problem Set 2, but here are two more triangles.
The triangle below has area and perimeter value of 24.

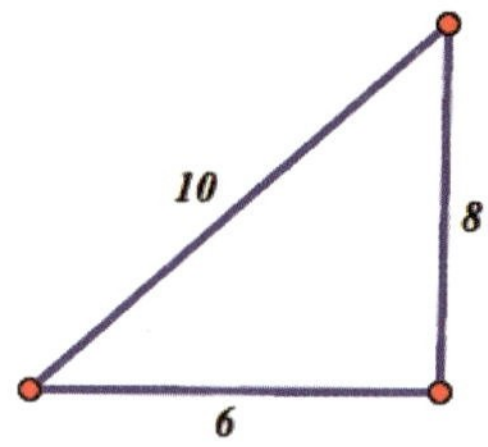

The triangle below has area and perimeter value of 30.

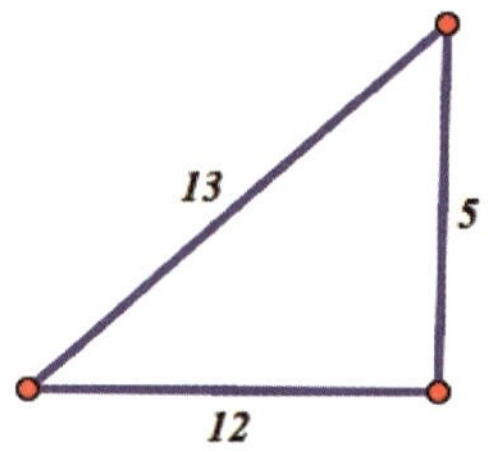

10. Pair 1: $a = 4, b = 4$
Pair 2: $a = 6, b = 3$
Pair 3: $a = 5, b = \frac{10}{3}$

11. A rectangle with sides a and b has area ab and perimeter $2(a + b)$. If the area is twice the perimeter (ignoring units!), then $ab = 4a + 4b$. This can be rewritten as $\frac{1}{4} = \frac{1}{a} + \frac{1}{b}$.

If the rectangle is a square, then $a = b = 8$. If one of the sides, say b, is greater than 8, then $a < 8$ for the equation to be true. Then to find all the solutions, we only need to check the possible values of a from 1 to 7. This gives us three rectangles: 5 by 20, 6 by 12, and 8 by 8.

An alternate algebraic approach would be to assume that $b \geqslant a$ and solve the equation to get

$b = \frac{4}{a-4}a$. Then we see that if $a = 5$, or $a = 6$, or $a = 8$, then $\frac{4}{a-4}$ is an integer and thus gives one of the required rectangles. But if $a > 8$ then $\frac{4}{a-4} < 1$, and $b < a$, which contradicts the assumption that $b \geqslant a$.

12. When $p(x) = x^4 - 2x^3 + 3$ is divided by $s(x) = (x-2)(x-5) = x^2 - 7x + 10$, the remainder is $q(x) = 125x - 247$.

Here is a table of some values of $p(x)$ and $q(x)$:

x	p(x)	q(x)
0	3	-247
1	2	-122
2	3	3
3	30	128
4	131	253
5	378	378

We observe that p and q have the same values when $x = 2$ and when $x = 5$! Why is this so? We can write an equation expressing the relationship between p and q:

$$p(x) = (x^2 + 5x + 25)(x-2)(x-5) + q(x)$$

The polynomial $(x^2 + 5x + 25)$ is the quotient of $p(x)$ divided by $s(x)$. We see from the equation that whenever $s(x) = 0$, i.e., when x is one of the roots of $s(x)$, the equation becomes $p(x) = 0 + q(x)$. This explains why the values in the table are the same for $x = 2$ and when $x = 5$.

13. If the quadratic polynomial $f(x) = 3x^2 - 10x + 21$ is divided by the quadratic function $s(x) = (x-3)(x+5)$, then the quotient will be a polynomial of degree zero (a constant) and the remainder $q(x)$ will be a polynomial of degree 1 (or 0). In this case the long division results in this equation:

$$f(x) = 3(x-3)(x+5) + (-16x + 66)$$

Thus $q(x) = -16x + 66$ is a linear function. By the same reasoning as in the previous problem, the functions $f(x)$ and $q(x)$ agree at the roots of $s(x)$, namely at $x = 3$, where $f(3) = q(3) = 18$, and when $x = -5$, where $f(-5) = q(-5) = 146$.

Moreover, the functions $f(x)$ and $q(x)$ *only* agree at these two points, for if $f(x) = q(x)$, then $3(x-3)(x+5) = 0$, so it must be true that $x = 3$ or

$x = 5$. Thus the graph of $q(x)$ is a secant line passing through the graph of $f(x)$ only when $x = 3$ and $x = -5$.

14. This problem is the same as before, with the same $f(x)$, except that both roots of the divisor polynomial are the same, so $s(x) = (x-3)^2$ has a double root.

Then following the same reasoning as before, we have

$$f(x) = 3(x-3)^2 + (8x-6)$$

Here $q(x) = 8x - 6$ is a linear function that agrees with $f(x)$ at $x = 3$, but is a very special linear function that agrees at this value because – as In Problem 14 – the two functions only agree when $x = 3$. So we conclude that the graph of the linear function is the tangent line of the graph of the quadratic function $f(x)$ at $x = 3$.

Another way to see that this is the tangent line is to consider nearby secants. For example, if the divisor function were $s(x) = (x-3)(x-3.0001)$, by the reasoning of Problem 14, the graph of the remainder would be a secant line intersecting the graph of $f(x)$ at $x = 3$ and $x = 3.0001$. The roots are so close, the figure would look the same; the secant would be very close to the tangent. So we can view the line in the graph below as the limit of secants.

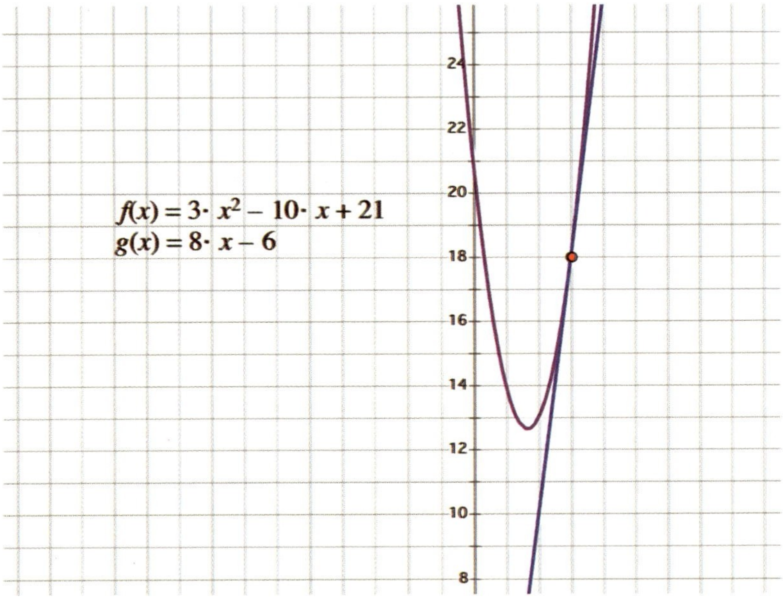

15. Using the same reasoning as before, we use long division to produce this equation:

$$x^3 = (x+4)(x-2)^2 + (12x - 16)$$

From the equation, we can see that $t(x) = 12x - 16$ agrees with x^3 when $x = 2$ and and also when $x = -4$. Because of the double root in the divisor, the graph of $t(x)$ is the line that is tangent to the graph of x^3 at $x = 2$.

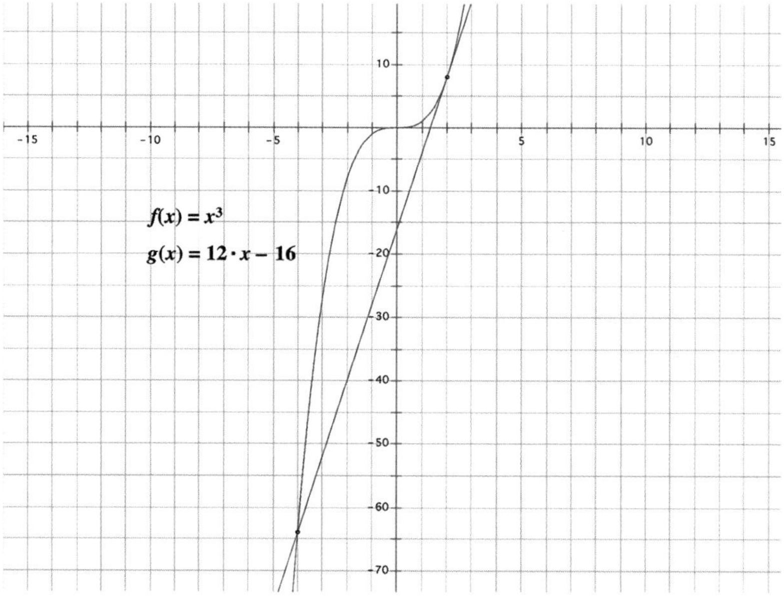

16. The sum does *not* stay the same as point P moves within a generic triangle.

Problem Set 4

1. Since this is an equilateral triangle, the line segment h bisects the side $\overline{CD}$ as in the picture below:

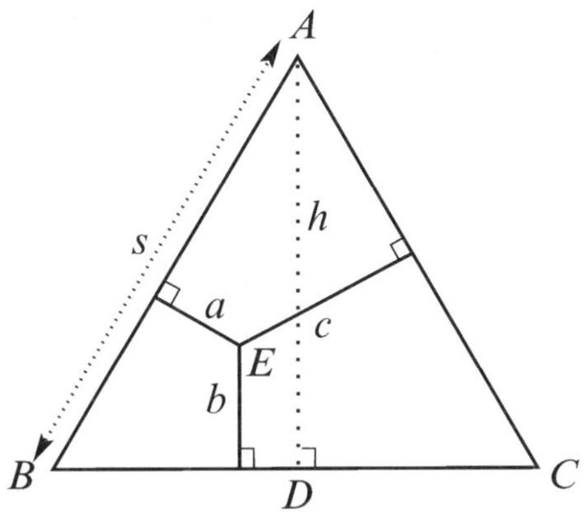

a. Side $\overline{BD}$ will measure $\frac{1}{2}s$. We have a right triangle and can apply the Pythagorean Theorem to find the height, h.

$$s^2 = \left(\frac{1}{2}s\right)^2 + h^2 \Rightarrow \sqrt{s^2 - \frac{1}{4}s^2} = h \Rightarrow \sqrt{\frac{3}{4}s^2} = h$$

$$h = \frac{\sqrt{3}}{2}s$$

Thus the area of the triangle is

$$A = \frac{1}{2} \cdot s \cdot \frac{\sqrt{3}}{2}s = \frac{\sqrt{3}}{4}s^2.$$

b. Look at the picture of the triangle above. Can you identify a triangle that has height a and base s? Yes! It is hiding a bit, but notice that $\triangle ABE$ is such a triangle!

c. Do what you did in part (b) to find triangles with heights b and c, respectively. $\triangle BEC$ and $\triangle ACE$ are such triangles. We can sum the area of all 3 triangles to get the area of the large triangle ($\triangle ABC$).
$A_{\triangle ABC} = \frac{1}{2}as + \frac{1}{2}bs + \frac{1}{2}cs = \frac{1}{2}s(a+b+c)$

Opener

Construct triangles. Check!

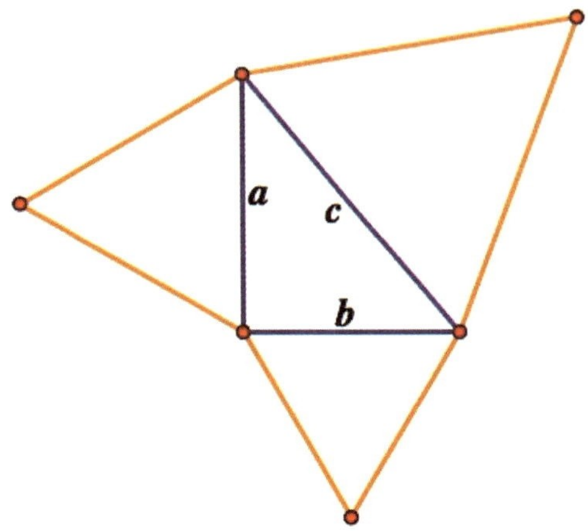

Let's take a look at one of our equilateral triangles. The altitude will bisect a side, making it easy to find the height.

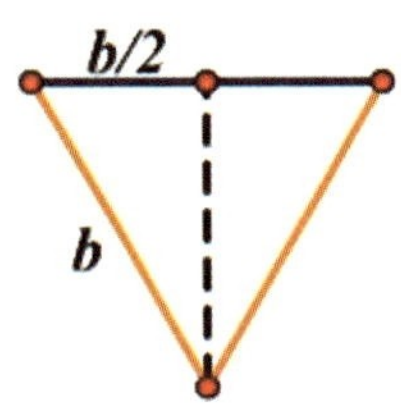

The height will be

$$h = \sqrt{b^2 - \frac{b^2}{4}} = \sqrt{\frac{3b^2}{4}} = \frac{b\sqrt{3}}{2},$$

so the area will be

$$A = \frac{1}{2}b\frac{b\sqrt{3}}{2} = \frac{\sqrt{3}}{4}b^2.$$

The areas of the equilateral triangles are:

$$\frac{\sqrt{3}}{4}a^2, \quad \frac{\sqrt{3}}{4}b^2, \quad \frac{\sqrt{3}}{4}c^2.$$

Notice that the sum of the areas of the smaller equilateral triangles is equal to the area of the largest equilateral triangle:

$$\frac{\sqrt{3}}{4}a^2 + \frac{\sqrt{3}}{4}b^2 = \frac{\sqrt{3}}{4}(a^2 + b^2) = \frac{\sqrt{3}}{4}c^2.$$

Opener, Part 2

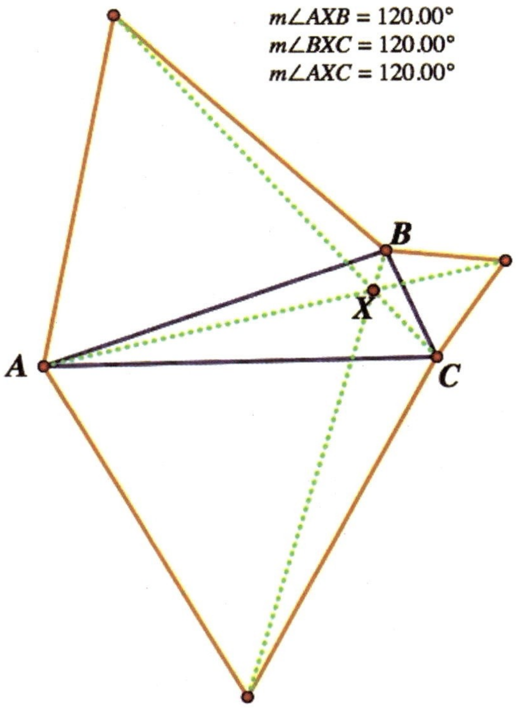

2. Go back to the solutions for Problem Set 1 . . .
   4×4: $\frac{1}{L} + \frac{1}{W} = \frac{1}{4} + \frac{1}{4} = \frac{1}{2}$
   3×6: $\frac{1}{L} + \frac{1}{W} = \frac{1}{3} + \frac{1}{6} = \frac{1}{2}$
   $8 \times \frac{8}{3}$: $\frac{1}{L} + \frac{1}{W} = \frac{1}{8} + \frac{3}{8} = \frac{1}{2}$
   $10 \times \frac{5}{2}$: $\frac{1}{L} + \frac{1}{W} = \frac{1}{10} + \frac{2}{5} = \frac{1}{2}$
   $7 \times \frac{14}{5}$: $\frac{1}{L} + \frac{1}{W} = \frac{1}{7} + \frac{5}{14} = \frac{1}{2}$
   $18 \times \frac{9}{4}$: $\frac{1}{L} + \frac{1}{W} = \frac{1}{18} + \frac{4}{9} = \frac{1}{2}$
   $20 \times \frac{20}{9}$: $\frac{1}{L} + \frac{1}{W} = \frac{1}{20} + \frac{9}{10} = \frac{1}{2}$
   And so on . . .

3. a. $\square ARON$ has area 300 and perimeter 70. Thus $R = \frac{300}{70} = \frac{30}{7}$, so we will plot the point $(20, \frac{30}{7})$. We will graph this point in part (b) with other points.

b. For a rectangle to be similar to $\square ARON$, the ratio of the side lengths must be the same as it is for $\square ARON$, which is $\frac{20}{15} = \frac{4}{3}$. Consider the following similar rectangles:

$l \times w$	P	A	$R = \frac{A}{P}$
4×3	14	12	$\frac{6}{7}$
8×6	28	48	$\frac{12}{7}$
16×12	56	192	$\frac{24}{7}$
24×18	84	432	$\frac{36}{7}$
$2 \times \frac{3}{2}$	7	3	$\frac{3}{7}$
40×30	140	1200	$\frac{60}{7}$

The graphs of the points (l, R) can be seen below:

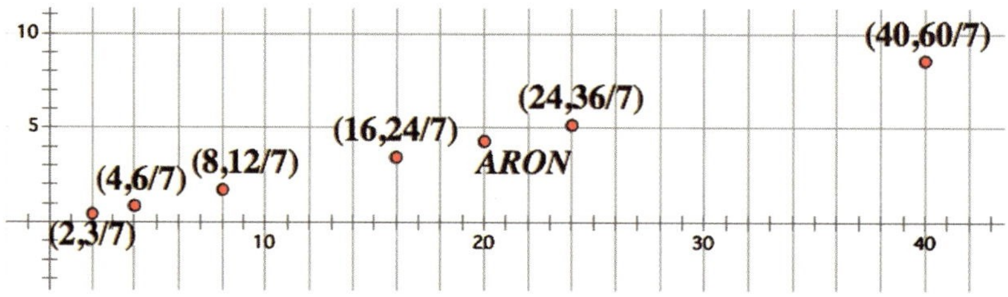

c. For a rectangle to be similar to $\square ARON$ and have the same value for both the perimeter and area, the following equations would need to be satisfied:

$2(l + w) = lw$ and $\frac{l}{w} = \frac{4}{3}$

The second equation gives $l = \frac{4}{3}w$. Plug this expression into the first equation.

$2 \cdot \frac{4w}{3} + 2w = \frac{4w}{3} \cdot w$

$8w + 6w = 4w^2 \Rightarrow 0 = 4w^2 - 14w \Rightarrow 0 = 2w(2w - 7) \Rightarrow w = 0, \frac{7}{2}$

If $w = \frac{7}{2}$, then $l = \frac{14}{3}$.

4. a. We want $lwh = 2(lw + lh + wh)$.

If Mimi's box is to be similar to Tom's box, then the ratios of the sides need to be equal. For Tom's box, let's use $l = 3$, $w = 4$, $h = 5$. We then have $\frac{l}{w} = \frac{3}{4} \Rightarrow l = \frac{3}{4}w$, and $\frac{h}{w} = \frac{5}{4} \Rightarrow h = \frac{5}{4}w$.

Now substitute these expressions for l and h into the equation.
$\frac{3}{4}w \cdot w \cdot \frac{5}{4}w = 2\left(\frac{3}{4}w \cdot w + \frac{3}{4}w \cdot \frac{5}{4}w + w \cdot \frac{5}{4}w\right)$
$\frac{15}{16}w^3 = \frac{3}{2}w^2 + \frac{15}{8}w^2 + \frac{5}{2}w^2$
$15w^3 = 24w^2 + 30w^2 + 40w^2 \Rightarrow 15w^3 = 94w^2$
$w^2(15w - 94) = 0 \Rightarrow w = 0, \frac{94}{15}$.
Clearly only $w = \frac{94}{15}$ makes sense in this case.
This gives $l = \frac{47}{10}$ and $h = \frac{47}{6}$.

b. $\frac{1}{T} + \frac{1}{O} + \frac{1}{M} = \frac{10}{47} + \frac{15}{94} + \frac{6}{47} = \frac{1}{2}$
Whoa! Why is that?!
Well, if a box is to have the surface area equal the value of the volume, then we have:
$lwh = 2(lw + lh + wh)$
Divide both sides of the equation by $2lwh$ to get $\frac{1}{2} = \frac{1}{h} + \frac{1}{w} + \frac{1}{l}$.
Bam!

5. **a.** The distance from $(20, 0)$ to $(10, 0)$ is $\sqrt{10^2 + 0^2} = 10$.
The distance from $(5, 0)$ to $(10, 0)$ is $\sqrt{5^2 + 0^2} = 5$.
Yes, it is twice as far from $(20, 0)$ as it is from $(5, 0)$.

b. The distance from $(20, 0)$ to $(-6, 8)$ is $\sqrt{26^2 + 8^2} = \sqrt{740} = 2\sqrt{185}$.
The distance from $(5, 0)$ to $(-6, 8)$ is $\sqrt{11^2 + 8^2} = \sqrt{185}$.
Yes, it is twice as far from $(20, 0)$ as it is from $(5, 0)$.

c. The distance from $(20, 0)$ to $(8, 6)$ is $\sqrt{12^2 + 6^2} = \sqrt{180} = 6\sqrt{5}$.
The distance from $(5, 0)$ to $(8, 6)$ is $\sqrt{3^2 + 6^2} = \sqrt{45} = 3\sqrt{5}$.
Yes, it is twice as far from $(20, 0)$ as it is from $(5, 0)$.

d. In order for the distance from some point (x, y) to be twice as far from $(20, 0)$ as it is from $(5, 0)$ we need to identify points (x, y) that satisfy the equation:
$\sqrt{(x-20)^2 + (y-0)^2} = 2\sqrt{(x-5)^2 + (y-0)^2}$
$(x-20)^2 + y^2 = 4((x-5)^2 + y^2)$
$x^2 - 40x + 400 + y^2 = 4x^2 - 40x + 100 + 4y^2$
$300 = 3x^2 + 3y^2 \Rightarrow 100 = x^2 + y^2$
All points (x, y) on the circle with center $(0, 0)$ and radius $r = 10$ are twice as far from the point $(20, 0)$ as from $(5, 0)$.

6. We want numbers x, y such that $xy = 49$ and $x+y$ is as small and as large as possible.
We can write $y = \frac{49}{x}$, which makes the sum $x+y = x + \frac{49}{x}$.
Notice that if $x = 7$, the sum is 14. If we make x larger or smaller than 7, the sum increases. Notice that as $x \to \infty$, the sum approaches ∞. The same is true as x approaches 0 on the right (i.e. we choose values near 0 that are positive). So for $x = 7$ we get a minimum sum of 14. We do not have a largest possible sum, but the sum will continue to increase as we choose positive values for x close to or out toward infinity. You can see this behavior is true when you look at the graph of the sum $x + \frac{49}{x}$, which is displayed below for your enjoyment!

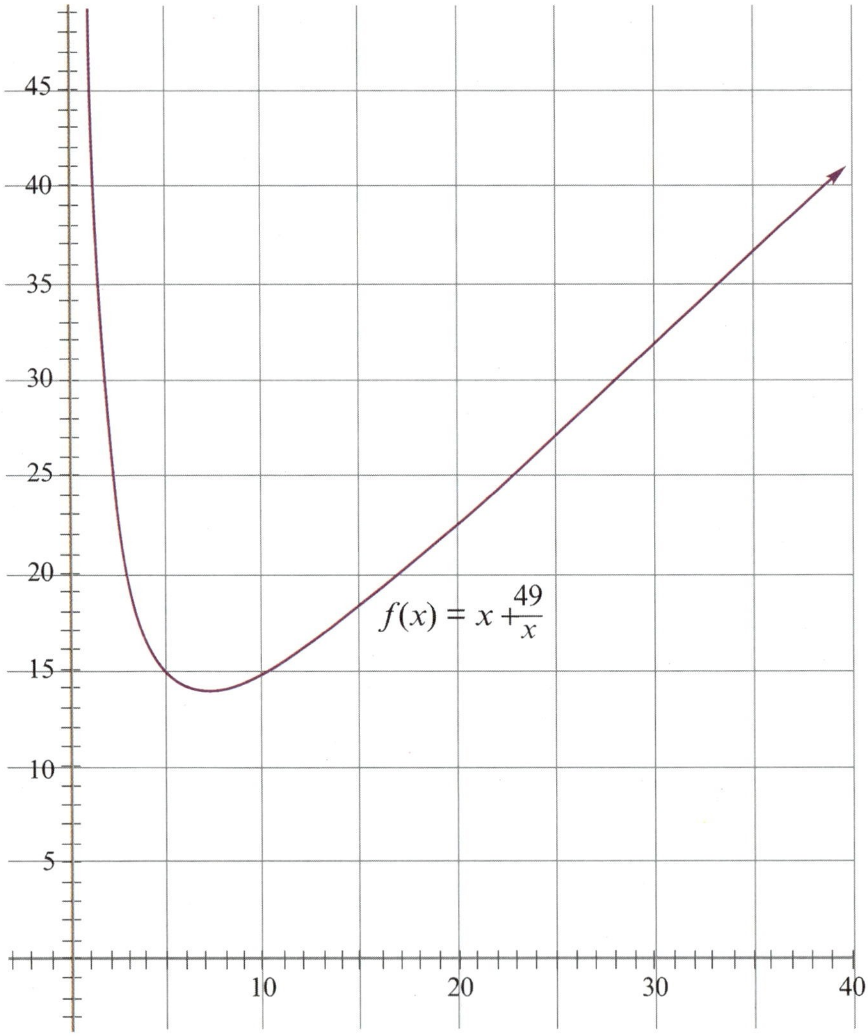

7. In each case, substitute $y = mx - 1$ into the equation of the circle.

   a. $x^2 + (2x-1)^2 = 1 \Rightarrow 5x^2 - 4x = 0 \Rightarrow x = 0, \frac{4}{5}$
   Now find y: $y = 2 \cdot \frac{4}{5} - 1 = \frac{3}{5}$, so $P = \left(\frac{4}{5}, \frac{3}{5}\right)$
   Hey now! 3-4-5 is a Pythagorean Triple!

   b. $x^2 + \left(\frac{3}{2}x - 1\right)^2 = 1 \Rightarrow 13x^2 - 12x = 0 \Rightarrow x = 0, \frac{12}{13}$
   Now find y: $y = \frac{3}{2} \cdot \frac{12}{13} - 1 = \frac{5}{13}$, so $P = \left(\frac{12}{13}, \frac{5}{13}\right)$.
   Hey hey now! 5-12-13 is a Pythagorean Triple!

   c. $x^2 + \left(\frac{4}{3}x - 1\right)^2 = 1 \Rightarrow 25x^2 - 24x = 0 \Rightarrow x = 0, \frac{24}{25}$
   Now find y: $y = \frac{4}{3} \cdot \frac{24}{25} - 1 = \frac{7}{25}$, so $P = \left(\frac{24}{25}, \frac{7}{25}\right)$.
   Hey hey hey! 7-24-25 is another PT!

   d. $x^2 + (4x-1)^2 = 1 \to 17x^2 - 8x = 0 \Rightarrow x = 0, \frac{8}{17}$
   Now find y: $y = 4 \cdot \frac{8}{17} - 1 = \frac{15}{17}$, so $P = \left(\frac{8}{17}, \frac{15}{17}\right)$.
   And yes, another PT: 8-15-17.

   e. $x^2 + (10x-1)^2 = 1 \Rightarrow 101x^2 - 20x = 0 \Rightarrow x = 0, \frac{20}{101}$
   Now find y: $y = 10 \cdot \frac{20}{101} - 1 = \frac{99}{101}$, so $P = \left(\frac{20}{101}, \frac{99}{101}\right)$
   20-99-101 is, of course, a PT.

   f. $x^2 + \left(\frac{a}{b}x - 1\right)^2 = 1 \Rightarrow x^2\left(1 + \frac{a^2}{b^2}\right) - \frac{2a}{b}x = 0$
   $\Rightarrow x^2\left(\frac{b^2+a^2}{b^2}\right) - \frac{2a}{b}x = 0 \Rightarrow x\left(\frac{b^2+a^2}{b^2}x - \frac{2a}{b}\right) = 0$
   $x = 0, \frac{2ab}{a^2+b^2}$.
   Now find y: $y = \frac{a}{b} \cdot \frac{2ab}{a^2+b^2} - 1 = \frac{2a^2}{a^2+b^2} - \frac{a^2+b^2}{a^2+b^2} = \frac{a^2-b^2}{a^2+b^2}$
   $P = \left(\frac{2ab}{a^2+b^2}, \frac{a^2-b^2}{a^2+b^2}\right)$

8. Start with the cubic polynomial $f(x) = x^3 - 6x^2 + 4x + 8$. We divide $f(x)$ by each of the quadratic polynomials below to find the remainder which will be a polynomial of degree 1 or 0. Then we graph and look for relationships.

   a. The quadratic $m(x) = (x-2)(x-6) = x^2 - 8x + 12$.

   $$f(x) = (x+2)m(x) + (8x - 16)$$

The cubic polynomial $(x + 2)m(x) = (x + 2)(x-2)(x-6)$ has three roots at the two roots of $m(x)$, which are $x = 2$ and $x = 6$, and also at $x = -2$, the root of the quotient polynomial $x + 2$. At these values of x, the cubic function $f(x)$ and the linear function $g(x) = 8x - 16$ will have the same value. These are the x-values for the three points of intersection of the graphs of the two functions.

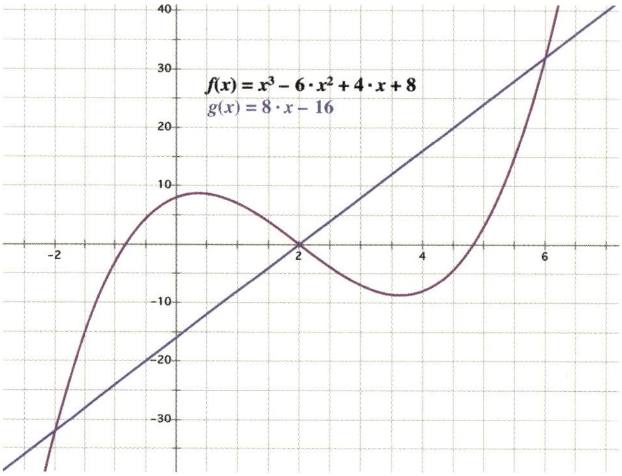

b. The quadratic $n(x) = (x - 2)(x - 5) = x^2 - 7x + 10$.

$$f(x) = (x + 1)n(x) + (x - 2)$$

As in (a), the polynomial $(x+1)n(x)$ has roots at $x = -1$, $x = 2$, and $x = 5$. Thus the graphs of $f(x)$ and $g(x) = x - 2$ intersect at these three values of x.

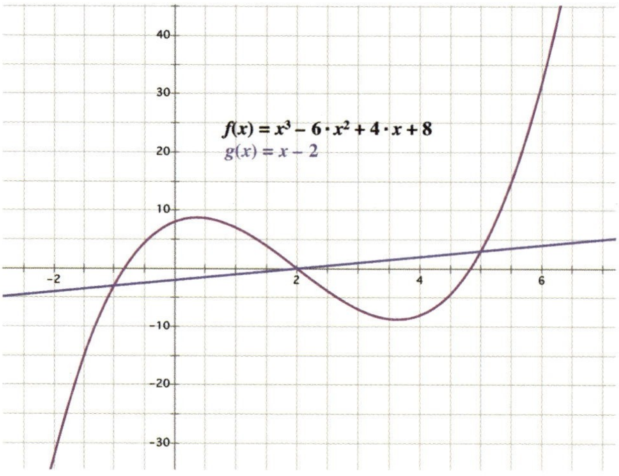

c. The quadratic $p(x) = (x-2)(x-4) = x^2 - 6x + 8$.

$$f(x) = xp(x) + (-4x + 8)$$

As in (a), the polynomial $xp(x)$ has roots at $x = 0$, $x = 2$, and $x = 4$. The graphs of $f(x)$ and $g(x) = x - 2$ intersect at these three values of x.

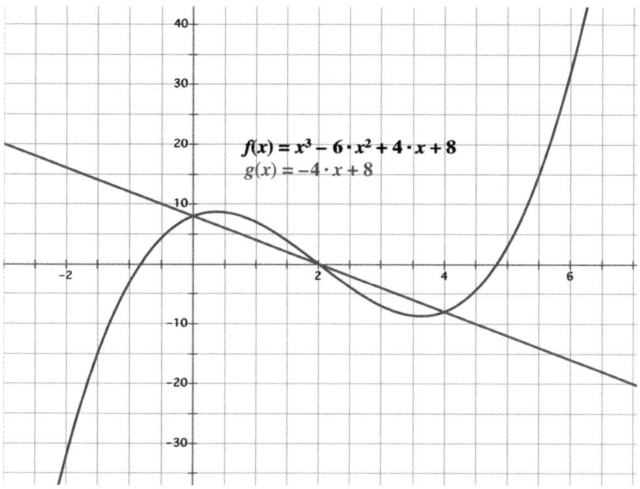

d. The quadratic $q(x) = (x-2)(x-3) = x^2 - 5x + 6$.

$$f(x) = (x-1)q(x) + (-7x + 14)$$

As in (a), the polynomial $(x-1)q(x)$ has roots at $x = 1$, $x = 2$, and $x = 3$. The graphs of $f(x)$ and $g(x) = x - 2$ intersect at these three values of x.

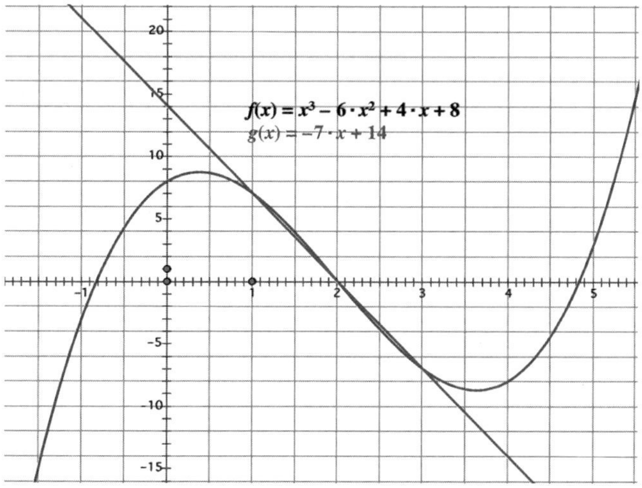

What do we notice? In each case, the sum of the three values of x where the graphs intersect is equal to 6! Why is this so? Notice that the cubic polynomial $(x-a)(x-b)(x-c) = x^3 - (a+b+c)x^2 + \dots$. This gives a relationship between the sum of the roots and the coefficient of x^2. Now observe the coefficients of x^2 on both sides of the equations in (a)-(d).

9. We continue with the same cubic polynomial $f(x) = x^3 - 6x^2 + 4x + 8$. Again we divide $f(x)$ by each of four quadratic polynomials to find the remainder which will be a polynomial of degree 1. We graph and look for relationships.

a. The quadratic $m(x) = (x-2)^2 = x^2 - 4x + 4$.

$$\begin{aligned} f(x) &= (x-2)m(x) + (-8x+16) \\ &= (x-2)^3 + (-8x+16) \end{aligned}$$

In this case, there is only one value $x = 2$ where the graphs intersect. The line which is the graph of $g(x)$ is tangent to the graph of $f(x)$ and this is also a point of inflection for the graph of $f(x)$. Once again, as before, the sum of the 3 intersection values of x equals $2+2+2=6$.

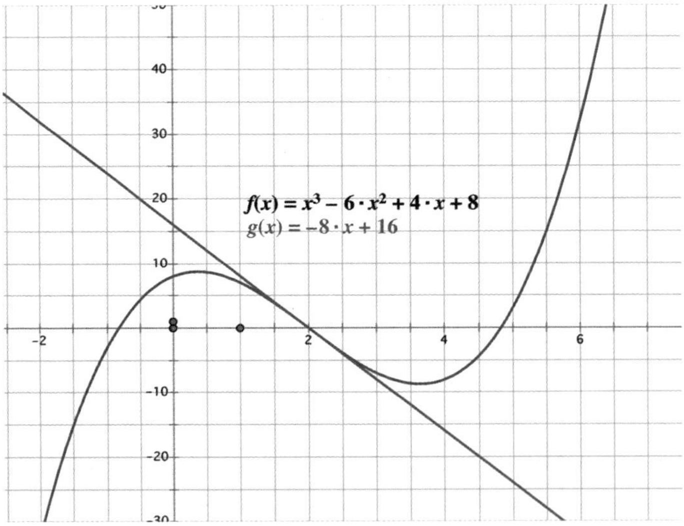

b. The quadratic $n(x) = (x-3)^2 = x^2 - 6x + 9$.

$$f(x) = xn(x) + (-5x+8) = x(x-3)^2 + (-5x+8)$$

In this case, the polynomial $xn(x)$ has a single root at $x = 0$ and a double root at $x = 3$. The graph of $g(x)$ is tangent to the graph of $f(x)$ at $x = 3$. This tangent line also intersects the cubic graph at $x = 0$. As expected, the sum of the 3 intersection values of x equals $0 + 3 + 3 = 6$.

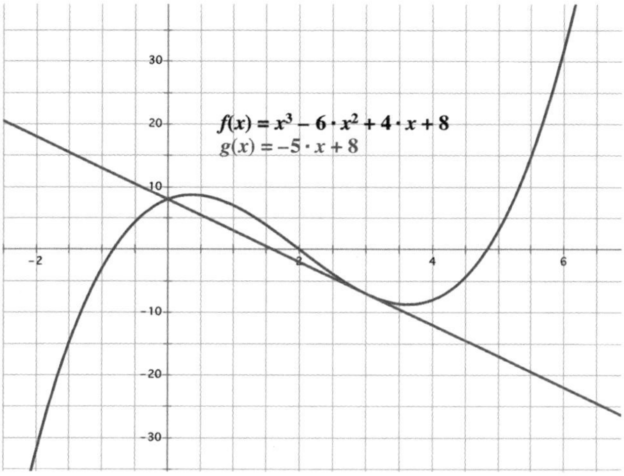

c. The quadratic $p(x) = (x-4)^2 = x^2 - 8x + 16$.

$$\begin{aligned} f(x) &= (x+2)p(x) + (4x - 24) \\ &= (x+2)(x-4)^2 + (4x - 24) \end{aligned}$$

In this case, the polynomial $(x + 2)p(x)$ has a single root at $x = -2$ and a double root at $x = 4$. The graph of $g(x)$ at $x = 4$ is tangent to the graph of $f(x)$. This tangent line also intersects the cubic graph at $x = -2$. Again the sum of the 3 intersection values of x equals $-2 + 4 + 4 = 6$.

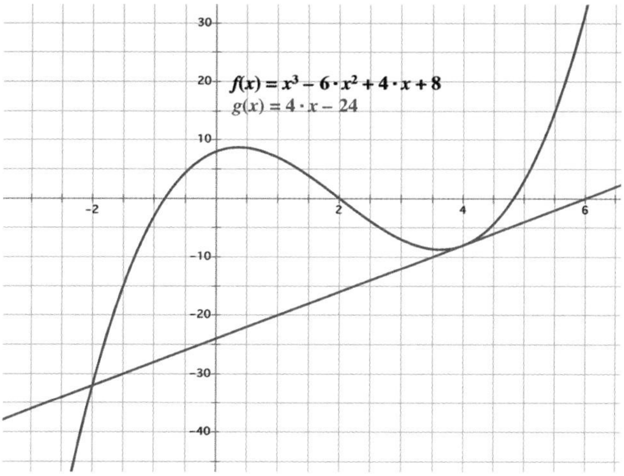

d. The quadratic $q(x) = (x-5)^2 = x^2 - 10x + 25$.

$$\begin{aligned} f(x) &= (x+4)q(x) + (19x - 92) \\ &= (x+4)(x-5)^2 + (19x - 92) \end{aligned}$$

In this case, the polynomial $(x+2)p(x)$ has a single root at $x = -4$ and a double root at $x = 5$. The graph of $g(x)$ at $x = 5$ is tangent to the graph of $f(x)$. This tangent line also intersects the cubic graph at $x = -4$. The sum of the 3 intersection values of x equals $-4 + 5 + 5 = 6$.

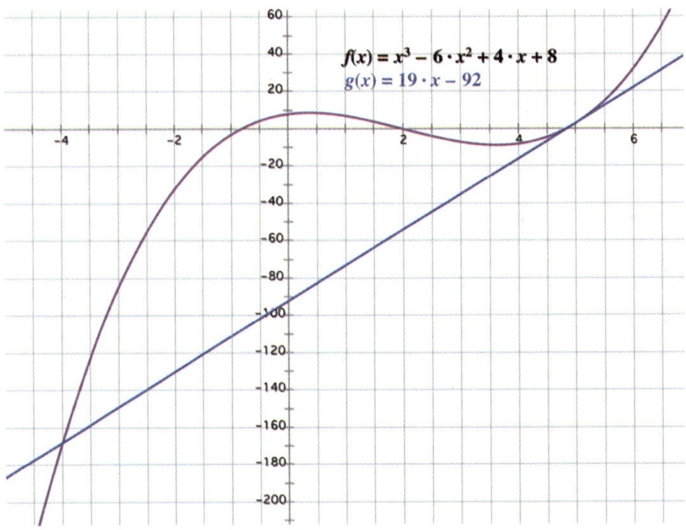

10. To find the tangent line to the graph of $f(x) = x^4$ at $x = 1$, we follow the same ideas from above and divide by $(x-1)^2$. The remainder will be a linear function whose graph is the tangent line we seek.

$$x^4 = (x^2 + 2x + 3)(x-1)^2 + (4x - 3)$$

The line $y = 4x - 3$ is tangent to the graph of x^4 at $x = 1$.

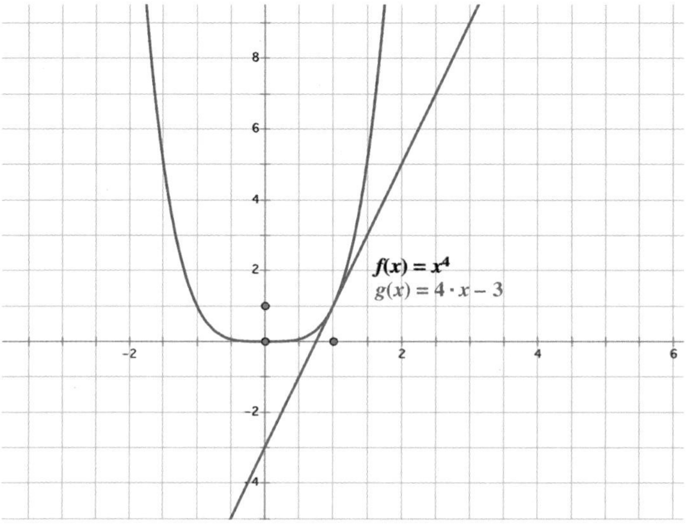

11. Given a polygon and a point P inside the polygon, construct perpendicular segments from P to each of the sides of the polygon – just as was done for equilateral triangles in the Opener. Is the sum of the lengths of the perpendicular segments a constant? This problem asks this question for 7 kinds of polygons.

One issue that comes up right away when the polygon has one or more obtuse interior angles is that the line through P that is perpendicular to a side may not intersect the segment that is the side. So we must decide whether or not to allow the foot of the perpendicular segment to lie on the line that is the side *extended*, not necessarily on the segment that is the side.

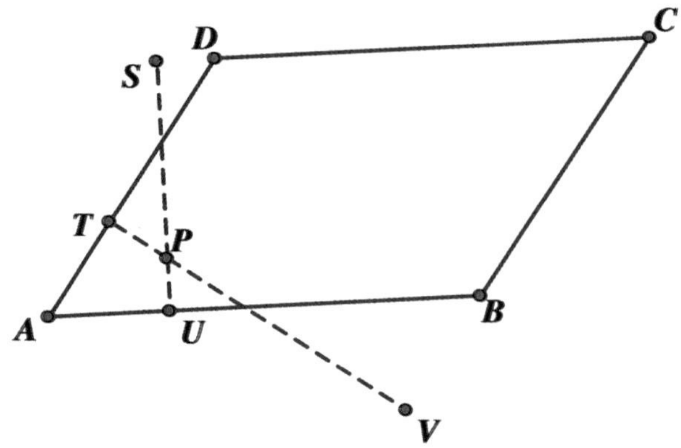

Since the sum will not exist if not all the segments exist, let's allow the end points of the perpendicular segments to lie on the extended sides.

The first four shapes are (a) square (b) rectangle (c) rhombus and (d) parallelogram. For squares, the feet of the perpendicular segments are on the sides and the sum is constant by the same area reasoning as for equilateral triangles. This area reasoning does not work for the other shapes, since the argument depends on the fact that the sides, which are bases of triangles, are all congruent. Nonetheless, the sum of the lengths is constant for all the cases (a)-(d). Here is the reasoning for parallelograms. Since all the shapes (a)-(d) are parallelograms, this proves the constancy for all four cases.

For a parallelogram ABCD and a point P inside, construct the perpendicular segments to each side (extended). Let PU be the segment from P to side AB and let PS be the segment from P to side CD. Then since these sides are parallel, both PU and PS are perpendicular to *each* of the sides. Thus PU and PS are both perpendicular to line AB, so both lines PU and PS are the same line. And P is in the interior of segment US. Thus the sum of the lengths $|UP| + |PS| = |US|$, which is the distance between the parallel lines AB and CD. This distance does not depend on P, In the same way, the sum of the lengths of the other two perpendicular segment is equal to the distance between the parallel lines BC and DA. Therefore, the sum of the four segment lengths equals the sum of the distances between the pairs of opposite sides of the parallelogram, and this is a constant.

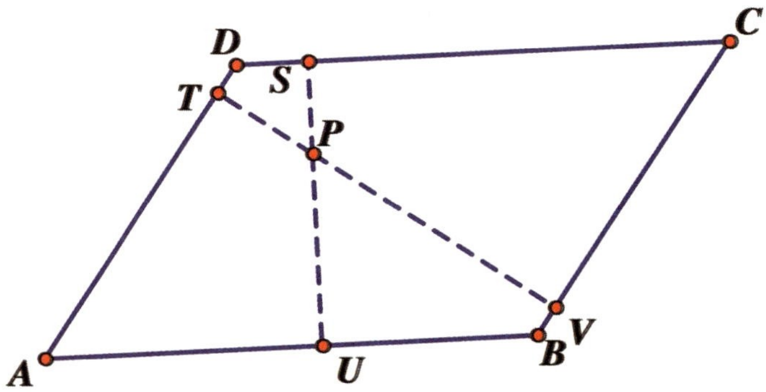

For case (e) the polygon is a regular pentagon ABCDE with side length s. In this case, one can construct the segments PA, PB, PC, PD, and PE which, along with the sides, dissect the pentagon into 5 triangles. The base of each triangle is a side of the regular pentagon, so each base length equals s. The altitudes of the triangles, h_1, h_2, h_3, h_4, h_5, are equal to the lengths to the five perpendicular segments from P to the sides. The sum of the areas of the triangles is

$$\frac{1}{2}(sh_1 + sh_2 + sh_3 + sh_4, +sh_5)$$
$$= \frac{s}{2}(h_1 + h_2 + h_3 + h_4 + h_5).$$

This is a constant times the sum of the lengths of the perpendicular segments. But this sum of areas of the triangles equals the area of the pentagon, which is a constant independent of P. So the sum of the lengths of the perpendicular segments is independent of P.

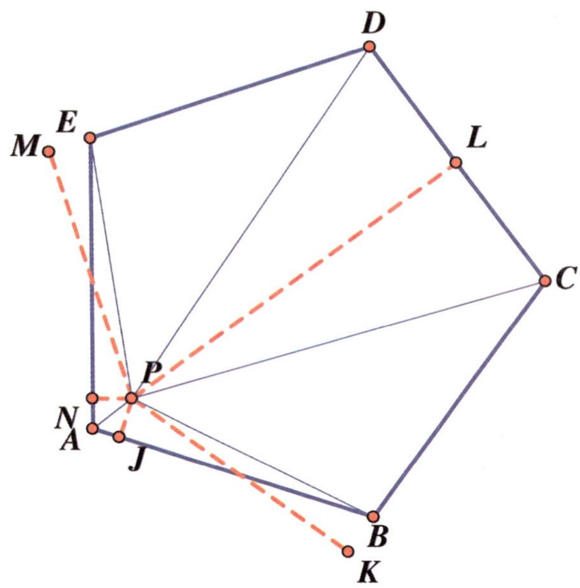

Case (e) of equilateral hexagons splits into two cases. If the hexagon is convex, the sum is constant for the same reasons as for the regular pentagon. But if the hexagon is not convex, for some points P the sum of the triangle areas is not equal to the area of the hexagon, so the sum of the lengths of the perpendicular segments is not the same for all P.

For case (f) of equiangular hexagons, the opposite sides of the hexagon are parallel, so the sum of the lengths of the perpendicular segments equals the sum of the three distances between pairs of opposite sides of the hexagon, just as in the case of parallelograms.

Problem Set 5

1. We want numbers x, y such that $xy = 81$ and $x + y$ is as small and as large as possible.
We can write $y = \frac{81}{x}$, which makes the sum $x + y = x + \frac{81}{x}$.
Notice that if $x = 9$, the sum is 18. If we make x larger or smaller than 9, the sum increases. Notice that as $x \to \infty$, the sum approaches ∞. The same is true as x approaches 0 on the right (i.e. we choose values near 0 that are positive). So for $x = 9$ we get a minimum sum of 18. We do not have a largest possible sum, but the sum will continue to increase as we choose positive values for x close to or out toward infinity. You can see this behavior is true when you look at the graph of the sum $x + \frac{81}{x}$, which is displayed below for your enjoyment!

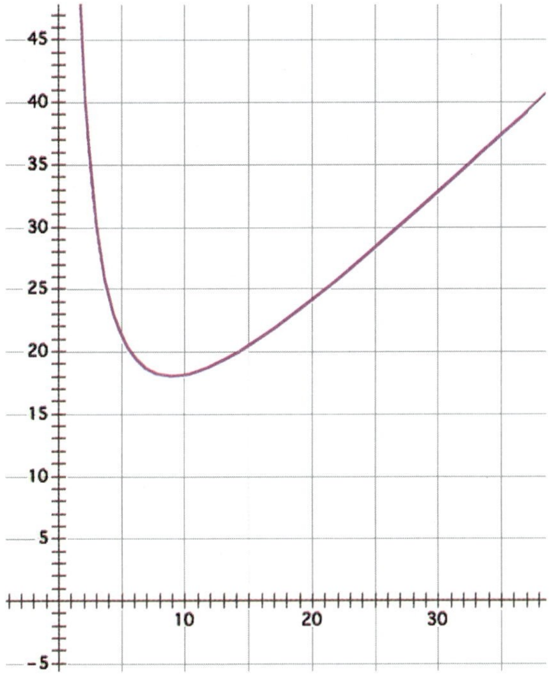

Opener

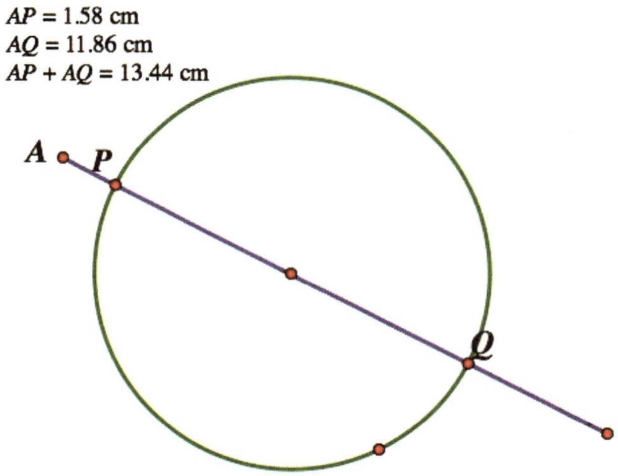

The maximum sum appears to be when the line segment passes through the center.

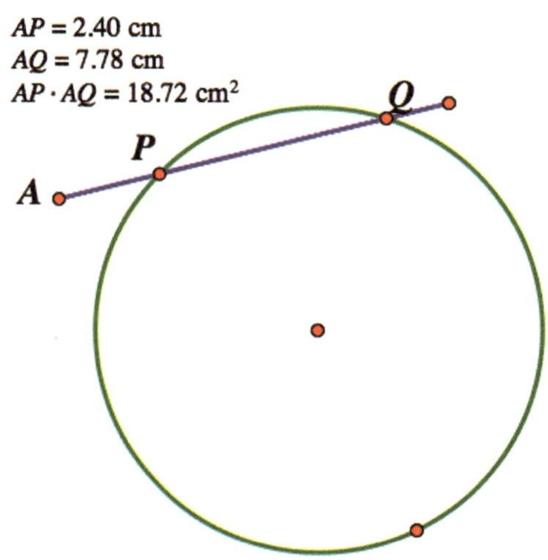

The product appears to be maximized when . . . Wait! The product is the same no matter where Q is located on the circle. Weird!

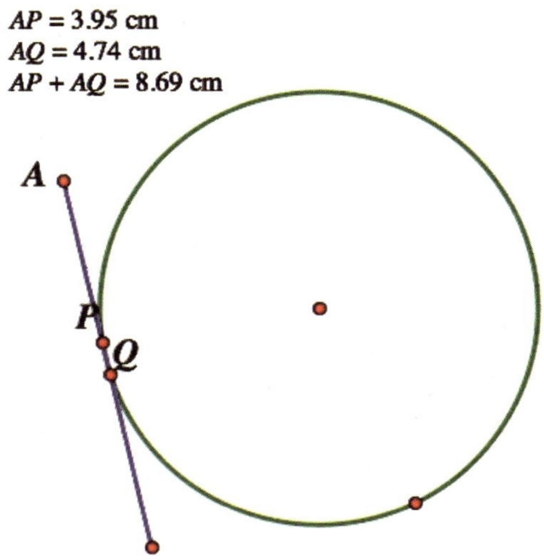

It looks like the sum will be minimized when P and Q coincide. Though, can't quite get that to happen with GSP . . .

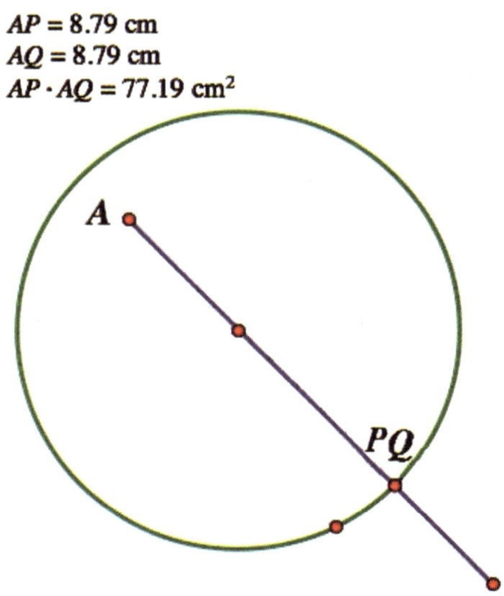

Now that A is inside the circle, P and Q coincide. The product appears to be maximized with passing through the center.

2. First of all, we want to be similar to a 7×11 rectangle, so we want $\frac{l}{w} = \frac{7}{11} \Rightarrow l = \frac{7}{11}w$ In addition, we want area to equal perimeter. Using ideas from previous problem sets, we know that $\frac{1}{l} + \frac{1}{w} = \frac{1}{2}$ in this scenario. So we have
$\frac{1}{(7/11)w} + \frac{1}{w} = \frac{1}{2}$
$\frac{11}{7w} + \frac{1}{w} = \frac{1}{2} \Rightarrow 22 + 14 = 7w$
$\frac{36}{7} = w$ and $\frac{36}{11} = l$
If you prefer to work with fractions, then you could have gotten a common denominator instead:
$\frac{11}{7w} + \frac{1}{w} = \frac{1}{2}$
$\frac{11}{7w} + \frac{7}{7w} = \frac{1}{2}$
$\frac{18}{7w} = \frac{1}{2} \Rightarrow 36 = 7w \Rightarrow \frac{36}{7} = w$
OR . . . if you are feeling clever! Multiply both sides by $2w$
$\frac{11}{7w} + \frac{1}{w} = \frac{1}{2}$
$\frac{22}{7} + 2 = w \Rightarrow \frac{36}{7} = w$
Then find l . . .

3. **a.** You can approach this problem in more than one way. Check it out!

Option 1:

We start with Todd's box, which is $5 \times 6 \times 20$. If Betul's box is similar then the dimensions of Todd's box is multiplied by a common scale factor, say s, to get Betul's box. Betul's box is $T \times O \times D$ and $T = 5s, O = 6s, D = 20s$
In addition, we know that in order to have the value of the volume equal the value of the surface area, $\frac{1}{T} + \frac{1}{O} + \frac{1}{D} = \frac{1}{2}$
Plug it all in!
$\frac{1}{5s} + \frac{1}{6s} + \frac{1}{20s} = \frac{1}{2}$
Multiply both sides by the LCD or multiply both sides by $2s$ or just do your favorite algebra!
$\frac{2}{5} + \frac{1}{3} + \frac{1}{10} = s$
$\frac{5}{6} = s$
This gives: $T = \frac{25}{6}, O = 5, D = \frac{50}{3}$

Option 2:
Let's set up some ratios, since we want Betul's box to be similar to Todd's.
$\frac{5}{6} = \frac{T}{O}$, $\frac{5}{20} = \frac{T}{D}$, and $\frac{6}{20} = \frac{O}{D}$
We can still use: $\frac{1}{T} + \frac{1}{O} + \frac{1}{D} = \frac{1}{2}$
We can use the ratio relationships to get our equation in terms of one variable. The variable is your choice, but here we will choose T.
$O = \frac{6T}{5}$, $D = 4T$. Substitute these expressions into the equation to get:
$\frac{1}{T} + \frac{1}{6T/5} + \frac{1}{4T} = \frac{1}{2}$
Now solve for T:
$\frac{1}{T}+\frac{5}{6T}+\frac{1}{4T} = \frac{1}{2} \Rightarrow 24+20+6 = 12T \Rightarrow T = \frac{25}{6}$
So $O = 5$ and $D = \frac{50}{3}$
Betul's box is $\frac{25}{6} \times 5 \times \frac{50}{3}$

b. We already know from previous problem sets that $\frac{1}{T} + \frac{1}{O} + \frac{1}{D} = \frac{1}{2}$
$\frac{6}{25} + \frac{1}{5} + \frac{3}{50} = \frac{12}{50} + \frac{10}{50} + \frac{3}{50} = \frac{25}{50} = \frac{1}{2}$

4. a. The distance from $(1, 0)$ to $(0, 3)$ is

$$\sqrt{(1-0)^2 + (0-3)^2} = \sqrt{10}.$$

The distance from $(9, 0)$ to $(0, 3)$ is

$$\sqrt{(9-0)^2 + (0-3)^2} = \sqrt{90} = 3\sqrt{10}.$$

Yes, $(0, 3)$ is three times as far from $(9, 0)$ as it is from $(1, 0)$.

b. The distance from $(1, 0)$ to $(7, 0)$ is

$$\sqrt{(1-7)^2 + (0-0)^2} = 6.$$

The distance from $(9, 0)$ to $(7, 0)$ is

$$\sqrt{(9-7)^2 + (0-0)^2} = 2.$$

No, $(7, 0)$ is not three times as far from $(9, 0)$ as it is from $(1, 0)$.

c. The distance from $(1, 0)$ to $(-2, 2)$ is

$$\sqrt{(1+2)^2 + (0-2)^2} = \sqrt{13}.$$

The distance from $(9, 0)$ to $(-2, 2)$ is

$$\sqrt{(9+2)^2 + (0-2)^2} = \sqrt{125} = 5\sqrt{5}.$$

No, $(-2, 2)$ is not three times as far from $(9, 0)$ as it is from $(1, 0)$.

d. The distance from $(1, 0)$ to $(-3, 0)$ is

$$\sqrt{(1+3)^2 + (0-0)^2} = 4.$$

The distance from $(9, 0)$ to $(-3, 0)$ is

$$\sqrt{(9+3)^2 + (0-0)^2} = 12.$$

Yes, $(-3, 0)$ is three times as far from $(9, 0)$ as it is from $(1, 0)$.

e. The distance from $(1, 0)$ to $(1, -5)$ is

$$\sqrt{(1-1)^2 + (0+5)^2} = 5.$$

The distance from $(9, 0)$ to $(1, -5)$ is

$$\sqrt{(9-1)^2 + (0+5)^2} = \sqrt{89}.$$

No, $(1, -5)$ is not three times as far from $(9, 0)$ as it is from $(1, 0)$.

f. For a point (x, y) to be three times as from from $(9, 0)$ as it is from $(1, 0)$, the following equation would have to be satisfied:

$$\sqrt{(9-x)^2 + (0-y)^2} = 3\sqrt{(1-x)^2 + (0-y)^2}$$

$$(9-x)^2 + y^2 = 9((1-x)^2 + y^2)$$

$$9^2 - 18x + x^2 + y^2 = 9 - 18x + 9x^2 + 9y^2$$

$$72 = 8x^2 + 8y^2 \qquad 9 = x^2 + y^2$$

All points (x, y) on the circle centered at $(0, 0)$ with radius 3 will be three times as far from the point $(9, 0)$ as from the point $(1, 0)$.

5. **a.** The graph of $\triangle SAM$ is below:

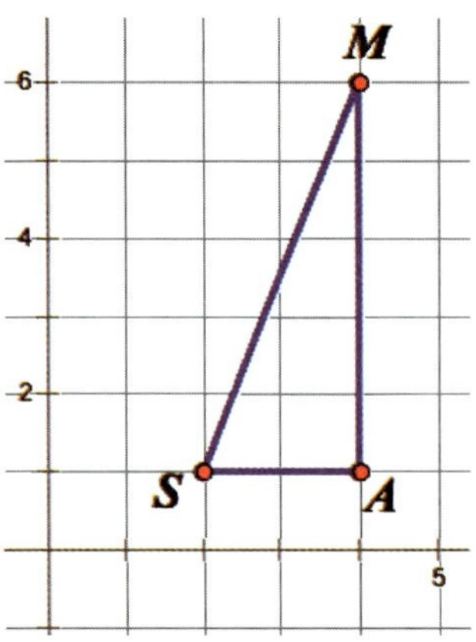

b. $\triangle SAM$ was rotated about the origin by 90° to get $\triangle S'A'M'$ pictured below:

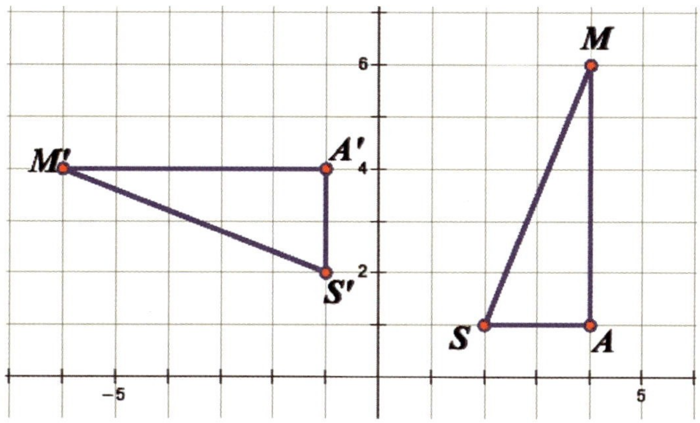

6. **a.** $s + m = (2 + i) + (4 + 6i) = 6 + 7i$

b. $s - m = (2 + i) - (4 + 6i) = -2 - 5i$

c. $3s + 3m = 3(2 + i) + 3(4 + 6i) = 3(s + m) = 18 + 21i$

d. $sm = (2+i) \cdot (4+6i) = 8 + 16i + 6i^2 = 2 + 16i$ (Remember that $i^2 = -1$)

e. $is = 2i + i^2 = -1 + 2i$

f. $im = 4i + 6i^2 = -6 + 4i$

7. When multiplying a complex number and its conjugate, you get $(a + bi)(a - bi) = a^2 - (bi)^2 = a^2 + b^2$

a. $\sqrt{(3 + 4i)(3 - 4i)} = \sqrt{3^2 + 4^2} = \sqrt{25} = 5$

b. $\sqrt{(5 - 12i)(5 + 12i)} = \sqrt{5^2 + 12^2} = \sqrt{169} = 13$

c. $\sqrt{(15 + 8i)(15 - 8i)} = \sqrt{15^2 + 8^2} = \sqrt{189} = 17$

d. $\sqrt{(x + yi)(x - yi)} = \sqrt{x^2 + y^2}$

8. **a.** $\sqrt{(3 - 0)^2 + (4 - 0)^2} = \sqrt{3^2 + 4^2} = 5$

b. $\sqrt{5^2 + (-12)^2} = 13$

c. $\sqrt{15^2 + 8^2} = 17$

d. $\sqrt{x^2 + y^2}$

Check out the previous problem! Same answers . . . why is that?

9. **a.** All points that are 5 units from the origin form a circle with center $(0, 0)$ and radius 5.

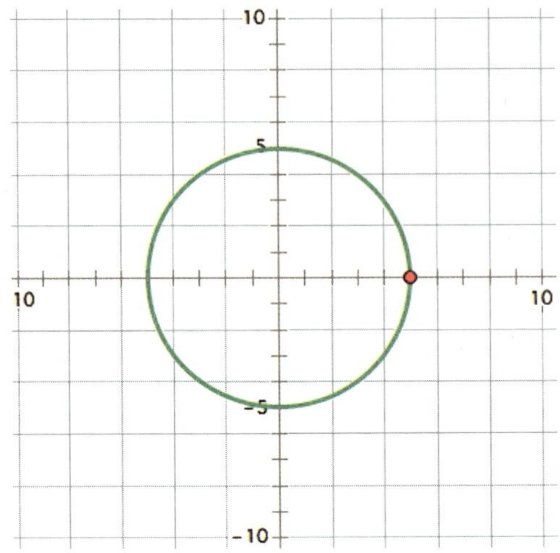

b. $x^2 + y^2 = 25$

10. We want complex numbers $a + bi$ such that $a^2 + b^2 = 5 \pm 5, \pm 5i, \pm 3 \pm 4i, \pm 4 \pm 3i$

11. $x = 2$ is a vertical line and points on the line are of the form $(2, y)$. When finding the distance from a point to the line, we want to use the same y-values, since the distance will be the horizontal distance from the point to the line $x = 2$.

a. The distance from $(5, 3)$ to $(-1, 3)$ is

$$\sqrt{(5+1)^2 + (3-3)^2} = \sqrt{36} = 6.$$

The distance from $(5, 3)$ to $x = 2$ is

$$\sqrt{(5-2)^2 + (3-3)^2} = \sqrt{9} = 3.$$

No, the distance from the point to the line is not twice as far as it is from the point to the point $(-1, 3)$; it is half as far.

b. The distance from $(0, 3)$ to $(-1, 3)$ is

$$\sqrt{(0+1)^2 + (3-3)^2} = 1.$$

The distance from $(0, 3)$ to $x = 2$ is

$$\sqrt{(0-2)^2 + (3-3)^2} = 2.$$

Yes, the distance from the point to the line is twice as far as it is from the point to the point $(-1, 3)$.

c. The distance from $(-2,4)$ to $(-1,3)$ is

$$\sqrt{(-2+1)^2+(4-3)^2}=\sqrt{2}.$$

The distance from $(-2,4)$ to $x=2$ is

$$\sqrt{(-2-2)^2+(4-4)^2}=\sqrt{16}=4$$

No, the distance from the point to the line is not twice as far as it is from the point to the point $(-1,3)$.

d. The distance from $(-3,2)$ to $(-1,3)$ is

$$\sqrt{(-3+1)^2+(2-3)^2}=\sqrt{5}.$$

The distance from $(-3,2)$ to $x=2$ is

$$\sqrt{(-3-2)^2+(2-2)^2}=\sqrt{25}=5$$

. No, the distance from the point to the line is not twice as far as it is from the point to the point $(-1,3)$.

e. The distance from $(-2,b)$ to $(-1,3)$ is

$$\sqrt{(-2+1)^2+(b-3)^2}=\sqrt{1+(b-3)^2}.$$

The distance from $(-2,b)$ to $x=2$ is

$$\sqrt{(-2-2)^2+(b-b)^2}=\sqrt{16}=4.$$

In order for the distance from $(-2,b)$ to $x=2$ to be twice as far as it is from $(-2,b)$ to $(-1,3)$ to, we would need points $(-2,b)$ to satisfy the equation:

$$4=2\sqrt{1+(b-3)^2}$$
$$2=\sqrt{10-6b+b^2}$$
$$4=10-6b+b^2$$
$$0=b^2-6b+6$$

$$\begin{aligned} b &= \frac{6\pm\sqrt{36-4(1)(6)}}{2} \\ &= \frac{6\pm\sqrt{60}}{2}=\frac{6\pm2\sqrt{15}}{2}=3\pm\sqrt{15} \end{aligned}$$

So the points $(-2,3+\sqrt{15})$ and $(-2,3-\sqrt{15})$ are twice as far from the line $x=2$ as they are from the point $(-1,3)$.

f. The distance from some point (x, y) to $(-1, 3)$ is $\sqrt{(x+1)^2 + (y-3)^2}$.
The distance from some point (x, y) to $x = 2$ is $\sqrt{(x-2)^2}$.
In order for the distance from (x, y) to $x = 2$ to be twice as far as it is from (x, y) to $(-1, 3)$ to, we would need points (x, y) to satisfy the equation:

$$\sqrt{(x-2)^2} = 2\sqrt{(x+1)^2 + (y-3)^2}$$

$$(x-2)^2 = 4((x+1)^2 + (y-3)^2)$$

$$0 = 3x^2 + 12x + 4y^2 - 24y + 36$$

$$-36 = 3(x^2 + 4x) + 4(y^2 - 6y)$$

$$-36 + 12 + 36 = 3(x^2 + 4x + 4) + 4(y^2 - 6y + 9)$$

$$12 = 3(x+2)^2 + 2(y-3)^2$$

$1 = \frac{(x+2)^2}{4} + \frac{(y-3)^2}{6}$, which is an ellipse with center $(-2, 3)$, major vertices $(-2, 3 \pm \sqrt{6})$, and minor vertices $(-2 \pm 2, 3) = (0, 3), (-4, 3)$. Thus, all points (x, y) on the ellipse will have distance twice as far from the line $x = 2$ as the point $(-1, 3)$.

12. Consider the graph of ZACK below:

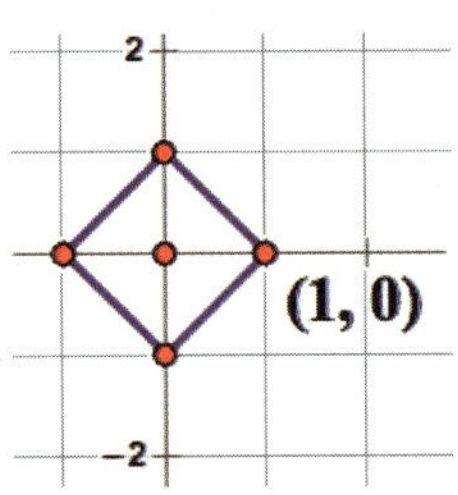

We can see that the remaining vertices will have coordinates $(0, 1)$, $(-1, 0)$, and $(0, -1)$.

13. We can start by factoring as a difference of squares.
$x^4 - 1 = (x^2 + 1)(x^2 - 1) = (x^2 - (-1))(x^2 - 1)$.
We can continue to factor using a difference of squares. Remember that $\sqrt{-1} = i$
$x^4 - 1 = (x^2+1)(x^2-1) = (x+i)(x-i)(x+1)(x-1) = 0$
Setting each factor equal to 0 and solving gives $x = \pm i, \pm 1$
Do you happen to see any relationship between the solutions to $x^4 - 1 = 0$ and square ZACK? Hmmm . . .

14. Consider the graph of DEB below:

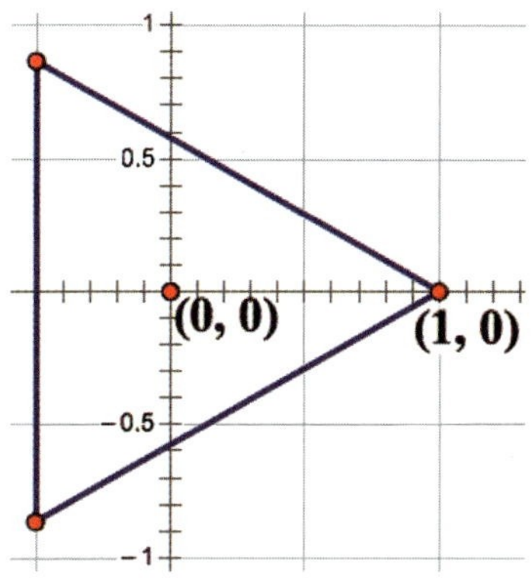

We can see that the x-axis bisects the angle at $(1, 0)$, so two 30-60-90 triangles are formed on top of each other. From what we know about these triangles or using trigonometry, we can determine that the remaining two vertices are $\left(-\frac{1}{2}, \frac{\sqrt{3}}{2}\right)$ and $\left(-\frac{1}{2}, -\frac{\sqrt{3}}{2}\right)$.

15. Begin factoring as a difference of cubes. If you cannot remember how to factor this, notice that $x = 1$ is a solution to $x^3 - 1 = 0$, which means that one of the factors of $x^3 - 1$ is $(x - 1)$. You can do long division, dividing $x^3 - 1$ by $(x - 1)$. Either way, you should get
$x^3 - 1 = (x - 1)(x^2 + x + 1) = 0$
Setting each factor equal to 0 and solving gives
$x = 1, \frac{-1 \pm \sqrt{-3}}{2}$
$x = 1, \frac{-1 \pm i\sqrt{3}}{2}$
$x = 1, -\frac{1}{2} \pm \frac{\sqrt{3}}{2}i$
Do you happen to see any relationship between the solutions to $x^3 - 1 = 0$ and triangle DEB? Hmmm . . .

16. Consider the graph of BUSHRA below:

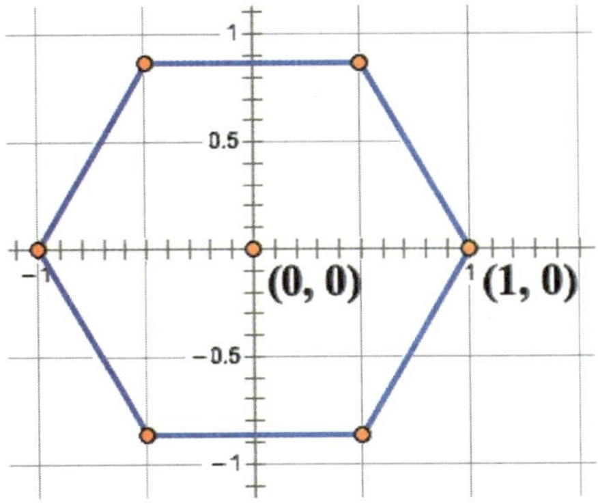

We can break the regular hexagon into smaller triangles and use properties of triangles and trigonometry to get the vertices: $(\pm 1, 0)$, $\left(\frac{1}{2}, \pm\frac{\sqrt{3}}{2}\right)$ and $\left(-\frac{1}{2}, \pm\frac{\sqrt{3}}{2}\right)$.

17. **Option 1:**
Begin factoring as a difference of squares:
$x^6 - 1 = (x^3 + 1)(x^3 - 1)$
Now factor using sum and difference of cubes:
$x^6 - 1 = (x^3 + 1)(x^3 - 1) = (x + 1)(x^2 - x + 1)(x - 1)(x^2 + x + 1) = 0$
Setting each factor equal to 0 and solving gives:
$x = \pm 1, -\frac{1}{2} \pm \frac{\sqrt{3}}{2}i, \frac{1 \pm \sqrt{-3}}{2}$
$x = \pm 1, -\frac{1}{2} \pm \frac{\sqrt{3}}{2}i, \frac{1}{2} \pm \frac{\sqrt{3}}{2}i$
Do you happen to see any relationship between the solutions to $x^6 - 1 = 0$ and hexagon BUSHRA? Hmmm . . .

Option 2:
Notice that ± 1 are solutions to the equation. Thus $(x + 1)(x - 1)$ are factors.
$(x + 1)(x - 1) = x^2 - 1$. You can perform long division, dividing $x^6 - 1$ by $(x^2 - 1)$ to get $x^4 + x^2 + 1$, which can be treated as a quadratic.

$x = \pm 1$ and $x^2 = \frac{-1 \pm \sqrt{-3}}{2}$. One can apply the Nth Root Theorem from trigonometry to solve $x^2 = \frac{-1 + \sqrt{-3}}{2}$ and $x^2 = \frac{-1 - \sqrt{-3}}{2}$ for x . . . This method might be painful and long.
Look back on the previous problems involving $x^3 - 1 = 0$ and $x^4 - 1 = 0$. What is the connection here? Looks like we are finding the 3rd, 4th, and 6th roots of 1. How could you check this? How does these connect to the regular 3, 4, and 6-gons?

18. $\left(\frac{\sqrt{2}}{2} + \frac{\sqrt{2}}{2}i\right)^2 = i$
$\left(\frac{\sqrt{2}}{2} + \frac{\sqrt{2}}{2}i\right)^3 = -\frac{\sqrt{2}}{2} + \frac{\sqrt{2}}{2}i$
$\left(\frac{\sqrt{2}}{2} + \frac{\sqrt{2}}{2}i\right)^4 = -1$
So $\left(\left(\frac{\sqrt{2}}{2} + \frac{\sqrt{2}}{2}i\right)^4\right)^2 = \left(\frac{\sqrt{2}}{2} + \frac{\sqrt{2}}{2}i\right)^8 = (-1)^2 = 1$
1 is pretty awesome!
Since $z^8 = 1$, $z^{101} = z^{8\cdot 12+5} = (z^8)^{12}z^5 = 1 \cdot z^5 = z^5 = -\frac{\sqrt{2}}{2} - \frac{\sqrt{2}}{2}i$

Problem Set 6

Opener

This is the graph of a circle of radius $\sqrt{65}$ with the center at the origin. We are looking for pairs integers (x, y)such that $x^2 + y^2 = 65$. When graphing in GSP, we can identify these points by seeing where the circle passes through integer points.

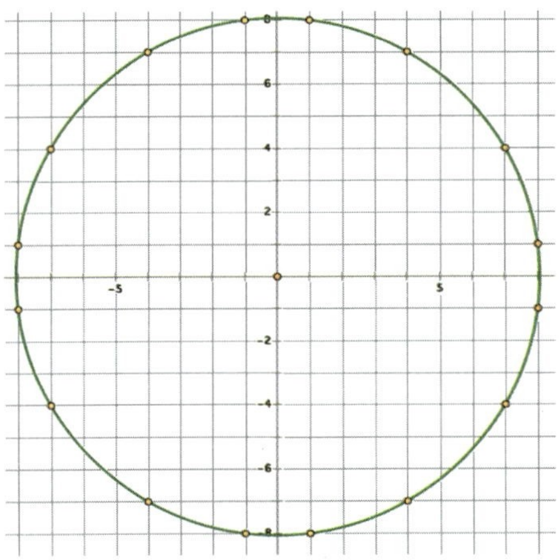

There are 16 of these points:
$(\pm1, 8), (\pm1, -8), (\pm8, 1), (\pm8, -1)$
$(\pm4, 7), (\pm4, -7), (\pm7, 4), (\pm7, -4)$

1. h^2 will be easiest to solve for in the first equation and then substitute into the second equation.
$h^2 = 13^2 - x^2$, which makes the second equation:
$13^2 - x^2 + (15 - x)^2 = 14^2$
$13^2 - x^2 + 15^2 - 30x + x^2 = 14^2$
$13^2 + 15^2 - 30x = 14^2$
$\frac{13^2+15^2-14^2}{30} = x = \frac{198}{30} = \frac{33}{5}$

Then $h^2 = 13^2 - \left(\frac{33}{5}\right)^2 = \frac{3136}{25}$, so $h = \pm\sqrt{\frac{3136}{25}} = \pm\frac{56}{5}$

Note also that you could have subtracted the original two equations from each other which takes you to an equivalent form of the second step from above:

$$\begin{array}{ll} h^2 + x^2 & = 13^2 \\ -(h^2 + (15-x)^2 & = 14^2) \\ \hline x^2 - (15-x)^2 & = 13^2 - 14^2 \end{array}$$

and so on . . .

2. Let's call the height of the triangle h. Suppose h breaks the base of the triangle up into two pieces: x (on the left) and $15 - x$ (on the right). Now we have two right triangles and can set up two equations using the Pythagorean Theorem.
$h^2 + x^2 = 13^2$ and $h^2 + (15-x)^2 = 14^2$
Hold on! These are the same two equations from Problem 1, and we already have the answers. Whoo-Hoo!
$x = \frac{33}{5}$ and $h = \frac{56}{5}$ (We only use the positive value since we are referring to a side length of a triangle.)
So the area is $A = \frac{1}{2}(15)\left(\frac{56}{5}\right) = 84$ and the perimeter is $P = 42$.

3. If the triangles are similar, then there is a common scale factor $s > 0$ applied to all sides of the triangle as w ell as the height. So the sides of the similar triangle are $13s, 14s, 15s$ with height $\frac{56}{5}s$.
You might be asking yourself: "Why is the height also multiplied by s? I need to be convinced."
In the previous problem we saw that $h = \sqrt{13^2 - x^2}$.
In the similar triangle our new h, say h' is going to be $h' = \sqrt{(13s)^2 - (xs)^2} = \sqrt{s^2(13^2 - x^2)} = s\sqrt{13^2 - x^2}$. So h also gets multiplied by the scale factor of s.
If area is equal to perimeter, we have the equation
$13s + 14s + 15s = \frac{1}{2}(15)\left(\frac{56}{5}s\right)$
$42s = 84s^2 \Rightarrow 0 = 2s^2 - s \Rightarrow s = 0, \frac{1}{2}$.
Use $s = \frac{1}{2}$, which gives side lengths $\frac{13}{2}, \frac{14}{2}, \frac{15}{2}$.

4. Set up two equations as in previous problems:
$x^2 + h^2 = 4^2$ and $(15-x)^2 + h^2 = 13^2$
Substitute $h^2 = 4^2 - x^2$ into the second equation:
$(15-x)^2 + 4^2 - x^2 = 13^2$
$15^2 - 30x + 4^2 = 13^2 \Rightarrow x = \frac{15^2+4^2-13^2}{30} = \frac{12}{5}$, which

gives $h = \frac{16}{5}$.
$A = \frac{1}{2}(15)\left(\frac{16}{5}\right) = 24$

5. **a.** $\bar{z} = 5 - 2i$

b. $\overline{w} = 3 + 4i$, so $w + \overline{w} = 3 - 4i + 3 + 4i = 6$, $w\overline{w} = (3 - 4i)(3 + 4i) = 3^2 + 4^2 = 25$

c. Want $v = a + bi$ such that $v + \bar{v} = 14$, so $v + \bar{v} = a + bi + a - bi = 2a = 14 \Rightarrow a = 7$
v is any complex number of the form $7 + bi$, where b is a real number.

d. We want $v\bar{v} = a^2 + b^2 = 65$. So we need to think about what sum of squares of two integers result in 65: ± 1 and ± 8, ± 7 and ± 4
Thus we could have $v = \pm 8 \pm i, \pm 1 \pm 8i, \pm 7 \pm 4i, \pm 4 \pm 7i$

e. Let's use what we found in parts (c) and (d). We already know that $(7 + bi) + (\overline{7 + bi}) = 14$ from part (c). We can narrow things down by using our results from part (d). We can see that $(7 \pm 4i)(\overline{7 \pm 4i}) = 65$. Thus $v = 7 \pm 4i$

So what did you discover? Notice that $z + \bar{z} = 2\text{Re}(z)$ (we are using the fancy $\text{Re}(z)$ to represent the real part of z), and $z\bar{z}$ is the square of the distance z is from the origin.

6. Hey! This is just using part (e) from the previous problem! So, "look it *up*!"

7. **a.** $|5+2i| = \sqrt{(5 + 2i)(5 - 2i)} = \sqrt{25 + 4} = \sqrt{29}$.
$|3 - 4i| = \sqrt{(3 - 4i)(3 + 4i)} = \sqrt{9 + 16} = \sqrt{25} = 5$

b. $|z| = \sqrt{z\bar{z}} = \sqrt{(a + bi)(a - bi)} = \sqrt{a^2 + b^2}$

c. In order for the magnitude of a complex number to be 0, we would need $a^2 + b^2 = 0 \Rightarrow$ both a and b need to be 0. Therefore the only complex number with magnitude 0 is $0 + 0i$
The magnitude of a complex number can never be negative. By definition of a complex number, $a + bi$, both a and b are real numbers. Since the magnitude of $a + bi$ is defined to be $a^2 + b^2$ and both a^2 and b^2 are greater than or equal to 0, it is not possible to have a negative magnitude for any complex number.

d. We can use previous ideas here. We want $|z| = \sqrt{z\bar{z}} = \sqrt{a^2 + b^2} = \sqrt{65} \Rightarrow a^2 + b^2 = 65$.
We already know this answer! z could be any one of the complex numbers:
$\pm 8 \pm i, \pm 1 \pm 8i, \pm 7 \pm 4i, \pm 4 \pm 7i$

8. We know that the complex numbers $\pm 8 \pm i$, $\pm 1 \pm 8i$, $\pm 7 \pm 4i$, $\pm 4 \pm 7i$ all have magnitude $\sqrt{65}$. There are 16 complex numbers here.

9. **a.** $|5+2i| = \sqrt{5^2+2^2} = \sqrt{29}$.

b. $(5+2i)^2 = 21+20i$. $|21+20i| = \sqrt{20^2+21^2} = 29$. Hey! Notice that $|(5+2i)^2| = (\sqrt{29})^2 = 29$ Is this always true? In other words, if $z = a+bi$, does $|z^2| = |z|^2$? Let's see . . .
$z^2 = (a^2 - b^2) + 2abi$, so

$$\begin{aligned} |z^2| &= \sqrt{(a^2-b^2)^2 + (2ab)^2} \\ &= \sqrt{a^4 - 2a^2b^2 + b^4 + 4a^2b^2} \\ &= \sqrt{a^4 + 2a^2b^2 + b^4} = \sqrt{(a^2+b^2)^2} \\ &= (\sqrt{a^2+b^2})^2 = |z|^2. \end{aligned}$$

Yes!

c. We want numbers a, b such that $a^2 + b^2 = 29^2$. Use what was found in part (b). Notice that $20^2+21^2 = 29^2$, so the Pythagorean Triple is 20, 21, 29.

10. We want numbers a, b such that $a^2 + b^2 = 65^2$. We know that $\pm 8 \pm i$, $\pm 1 \pm 8i$, $\pm 7 \pm 4i$, $\pm 4 \pm 7i$ all have magnitude $\sqrt{65}$, so we can square any of them to get a number with magnitude 65, which can be used to determine a Pythagorean Triple. However, when doing this, be aware of the complex numbers that may result from squaring.
For example, $(8+i)^2 = 63+16i$, so our Pythagorean Triple would be 16, 63, 65.
However, $(1+8i)^2 = -63+16i$. This number has a magnitude of 65, but we cannot use a negative number as part of a Pythagorean Triple. But, we can see that the triple is 63, 16, 65.
Similarly for $7+4i$ vs. $-4+7i$
$(7+4i)^2 = 33+56i$ and $(-4+7i)^2 = -33-56i$. Both have magnitude 65, but the Pythagorean Triple is 33, 56, 65.

11. 12-10-3
15-6-4

12. $V = \frac{1}{360}$ and $SA = \frac{5}{36}$

13. Under the transformation rule $(x, y) \mapsto (5x - 2y, 2x - 5y)$, these points map thusly:

a. $P(1, 0) \mapsto (5, 2)$
b. $Q(1, 0) \mapsto (21, 20)$
c. $R(21, 20) \mapsto (65, 142)$
d. $S(65, 142) \mapsto (41, 840)$

Notice that each point in the list (b)-(d) is the image of the preceding point. Also, there seems to be some connection between these points and the complex numbers of Problem 9!

14. Yes. If $z = x + i0$, then $|z| = \sqrt{x^2 + 0^2} = |x|$, the usual absolute value of x.

15. **a.** Both primitive Pythagorean triples with $c = 65$ were found in the solution for Problem 10. Two triples are 33, 56, 65 and 16, 63, 65.

b. We can use the same method for the case $c = 85$. We begin by finding ways to express 85 as the sum of two squares. By checking all the squares less than 85, we find $49 + 36 = 7^2 + 6^2 = 85$ and $81 + 4 = 9^2 + 2^2 = 85$.

Expressing the same equations with complex numbers, if we set $z = 7 + 6i$ and $w = 9 + 2i$, then $|z|^2 = 7^2 + 6^2 = 85$ and $|w|^2 = 9^2 + 2^2 = 85$. Since $|z|^2 = 85$, then $|z^2|^2 = |z|^4 = 85^2$. But $z^2 = 13 + 84i$, so $85^2 = 13^2 + 84^2$.

In the same way $|w^2|^2 = 85^2$. Since $w^2 = 77 + 36i$ then $85^2 = 77^2 + 36^2$.

Thus two triples are 13, 84, 85 and 36, 77, 85.

One more thing! Rather than finding sums of squares that equal 65 and 85 by checking cases, there is a method using complex factors.

For part (a) we use the integer factorization $65 = 5 \times 13$,and the complex factorizations $5 = (1 + 2i)(1 - 2i) = |1 + 2i|^2$ and $13 = (2 + 3i)(2 - 3i) = |2 + 3i|^2$.

Since $|1+2i|^2 = 5$ and $|2+3i|^2 = 13$, then $|(1+2i)(2+3i)|^2 = 5 \times 13 = 65$. If we let $z = (1 + 2i)(2 + 3i) = -4 + 7i$, then $|z|^2 = 65$. So we get $4^2 + 7^2 = 65$, as we did before.

To get the other case, take the conjugate of one of the complex numbers. Let $w = (1 + 2i)(2 - 3i) = 8 + i$, also with $|w|^2 = 65$. This produces the other sum $8^2 + 1^2 = 65$.

For part (b), we have $85 = 5 \times 17, 5 = |1+2i|^2$, and $17 = |1+4i|^2$. Then we choose z to be the product $z = (1+2i)(1+4i) = -7+6i$. This implies $7^2+6^2 = 85$. If we choose $w = (1+2i)(1-4i) = 9-2i$. Then $|w|^2 = 9^2 + 2^2 = 85$, giving the second way of writing 85 as the sum of two squares.

16. If a, b, c, is a Pythagorean triple with the leg a being an even number, then the area of the triangle is the integer $ab/2$. Such triples provide many examples of right triangles with integer perimeter and integer area.

To produce an example of such a triangle that is not a right triangle, suppose a', b', c', is a second Pythagorean triple, with the leg a' an even number.

If it happens that $a = a'$, then by adjoining these two triangles along the sides with this length, then the triangle so constructed is a triangle with perimeter, $b+b'+c+c'$ and area $(ab+a'b')/2 = ab/2 + ab'/2$, both of which are integers.

If these two side lengths are not equal, then we can replace the two right triangles with similar right triangles so that the side lengths are equal.

Let the first triangle have sides aa', ba', ca' and the second triangle have sides $a'a$, $b'a$, $c'a$. These are two right triangles each with a side aa', an even integer. So combining the triangles as before produces a triangle – not a right triangle – with integer perimeter and area.

The sides of this triangle may have a common divisor, so there may be a smaller similar triangle that also has integer perimeter and area.

Problem Set 7

Opener

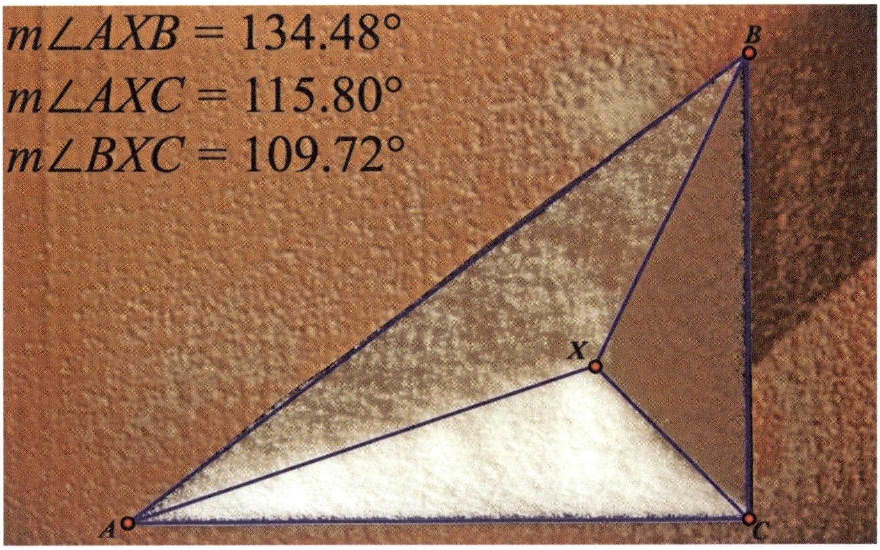

We can see that it is not the 120° point, but this is a 3D shape, so that makes sense. Actually, the angles at X all look like right angles. So we might just have a tetrahedron with all right triangles . . .

Hey! Looks what happens when we use X as the center of a circle and the circle touches the sides of the base triangle! And . . . the ridges appear to bisect the angles at the three base vertices . . .

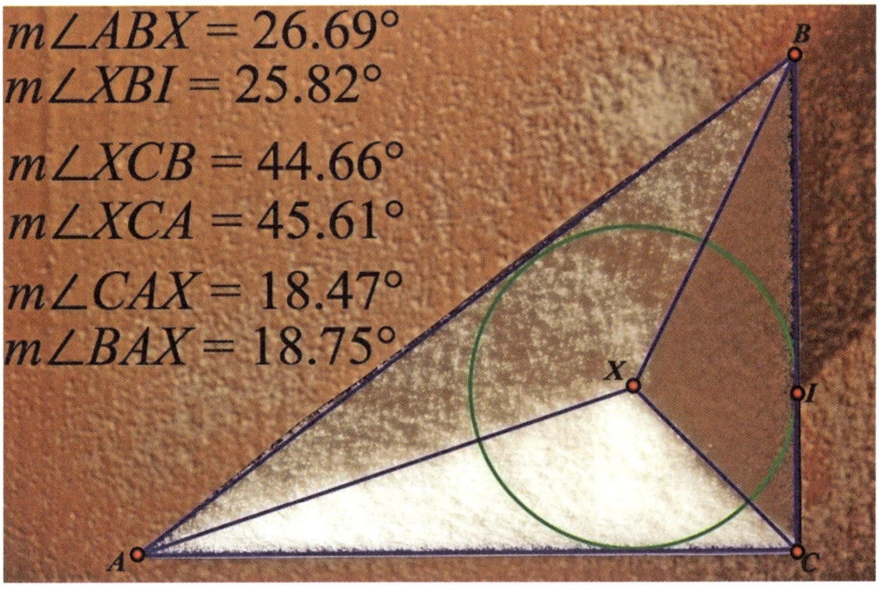

1. Let h be the height of the triangle. The line segment h divides up the base into two smaller line segments: $20 - x$ and 20
We can set up a system of two equations with two unknown quantities, x and h, as we did in the previous problem set.
(1) $x^2 + h^2 = 7^2$
(2) $(20 - x)^2 + h^2 = 15^2$
Solve for h^2 in (1): $h^2 = 7^2 - x^2$
Substitute the expression into (2): $(20 - x)^2 + 7^2 - x^2 = 15^2$
Solve for x: $20^2 - 40x_x^2 + 7^2 - x^2 = 15^2 \Rightarrow x = \frac{20^2+7^2-15^2}{40}$
$x = \frac{28}{5}, h = \frac{21}{5}$
$A = \frac{1}{2} \cdot 20 \cdot \frac{21}{5} = 42$
Notice that the perimeter is also 42 (Don't forget your towel!)! Whoa!

2. We've got this process down pat!
$x^2 + h^2 = 6^2$ and $(29 - x)^2 + h^2 = 25^2$
$h^2 = 6^2 - x^2$ Plug into the second equation to get:
$(29 - x)^2 + 6^2 - x^2 = 25^2 \Rightarrow 29^2 - 58x + 6^2 = 25^2$
$x = \frac{29^2+6^2-25^2}{58} = \frac{126}{29} \Rightarrow h = \frac{120}{29}$
$A = \frac{1}{2} \cdot 29 \cdot \frac{120}{29} = 60$
Notice that the perimeter is also 60! Whoa again!

3. The radius is about 2 units.

4. **a.** Look at the figure below:

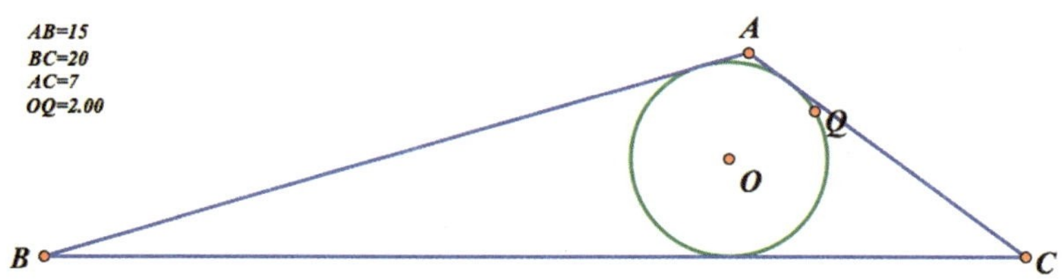

Notice that $\triangle BOC$ has $A = \frac{1}{2}(20)(2) = 20$. *Boom!*

b. Break $\triangle ABC$ up into 3 smaller triangles where each of these triangles have a height corresponding to the radius of the inscribed circle. $A_{\triangle ABC} = \frac{1}{2}(20(2) + \frac{1}{2}(15)(2) + \frac{1}{2}(7)(2) = \frac{1}{2}(2)(20 + 15 + 7) = 42$
Hey! Notice that we have $\frac{1}{2}Pr$ for area. Nice!

5. **a.** The graph for parts (a), (b), and (c) will be together in the graph below in part (c).

b. $s' = is = -1 + 2i$, $a' = ia = -1 + 4i$, $m' = im = -6 + 4i$

c. $s'' = i^2 s = -1s = -2 - i$, $a'' = i^2 a = -1a = -4 - i$, $m'' = i^2 m = -1m = -4 - 6i$

d. $s''' = i^3 s = -is = 1 - 2i$, $a''' = i^3 a = -ia = 1 - 4i$, $m''' = i^3 m = -im = 6 - 4i$

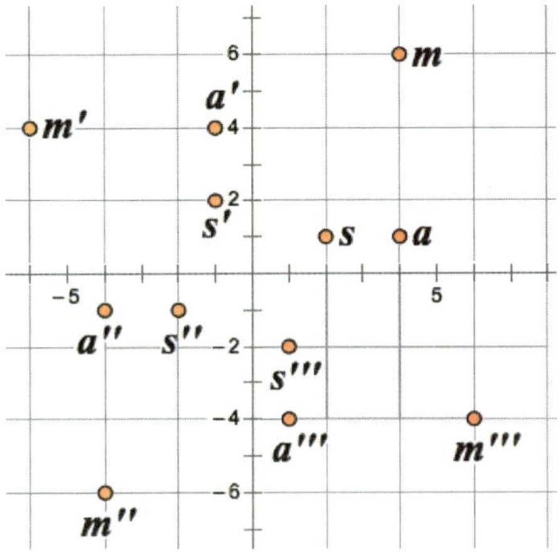

If we multiplied by i four times, then we would get our original points back, since $i^4 = 1$. Fives times is equivalent to multiplying by i, since $i^5 = i$. For i^{13}, you can divide 13 by 4 which has a remainder of 1, so $i^{13} = i$, so multiplying by i thirteen times is equivalent to multiplying by i once (i.e. $i^{13} = (i^4)^3(i)^1 = (1)(i) = i$). You can use modular arithmetic to see this too: $13 \text{mod}(4) = 1$. Similarly, for i^{101}, this is equivalent to multiplying by i, since the remainder when dividing 101 by 4 is 1.

6. The graph for all parts of this problem are displayed in a single graph in part (e)
   a. $|z| = \sqrt{2}$
   b. $z^2 = 2i$, $|z^2| = 2$
   c. $z^3 = -2 + 2i$, $|z^3| = 2\sqrt{2}$
   d. $z^4 = -4$, $|z^4| = 4$
   e. $z^5 = -4 - 4i$, $|z^5| = 4\sqrt{2}$
   Notice that each time we raise z to a power, the new magnitude gets multiplied by the magnitude of z.

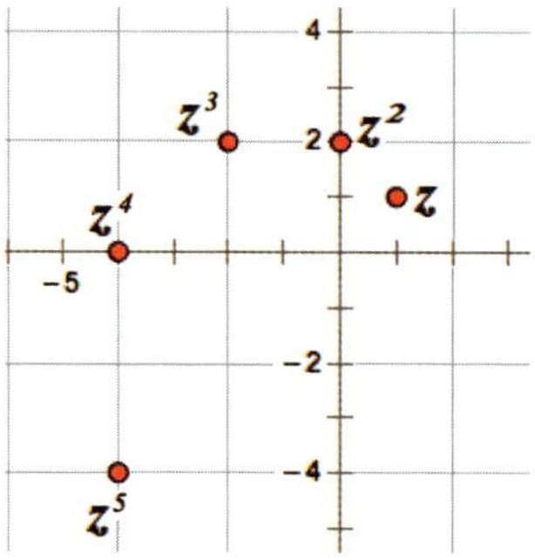

   This is an *Archimedean Spiral*, so feel free to look that up!

7. The points are getting further away and spiraling out away from Craig - maybe the powers of z don't like Craig.
   For each iteration the distance from Craig gets multiplied by the factor $\sqrt{2}$ (magnitude of z) and rotates in the counterclockwise direction by $45°$ (angle z forms with the positive x-axis). So on the kth iteration, the point z^k will be a distance of $(\sqrt{2})^k$ from Craig and will form a $45k°$ angle with the positive x-axis.

8. Remember what we discovered from Day 6: $|z|^2 = |z^2| \ldots$
   a. $|z| = \sqrt{3^2 + 2^2} = \sqrt{13}$
   b. $|z^2| = (\sqrt{13})^2 = 13$
   c. $z^3 = -9 + 46i$, $|z^3| = \sqrt{9^2 + 46^2} = \sqrt{2197} = 13\sqrt{13}$.
   Hey! It looks like $|z^3| = |z|^3$
   d. $|z^4| = |(z^2)^2| = |z^2|^2 = |z|^4$, so $|z^4| = (\sqrt{13})^4 = 13^2 = 169$

9. **a.** $z = \frac{3}{5} + \frac{4}{5}i$
b. $z = -\frac{7}{25} + \frac{4}{25}i$
c. $z = -\frac{117}{125} + \frac{44}{125}i$
d. $z = -\frac{527}{625} - \frac{336}{625}i$
e. $z = -\frac{237}{3125} - \frac{3116}{3125}i$

See the graph below for the plotted points (note that each tic mark is 1 unit):

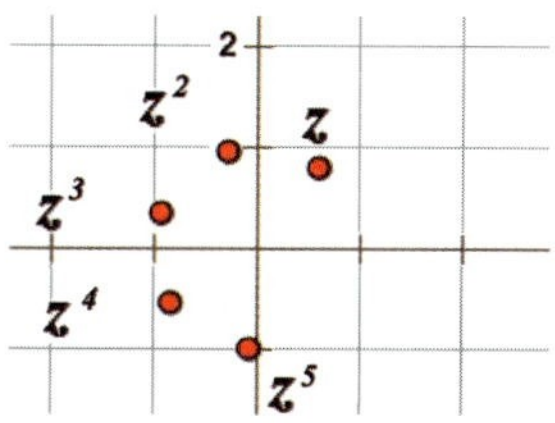

10. Unlike Craig, the points are remaining the same distance away from Mahen (1 unit). I guess they are indifferent to Mahen. This is because the magnitude of z in this case is 1. So, although the points behave the same here as they did with Craig, multiplying by 1 each time does not effect the distance. The points do circle around Mahen, though.
So, for each iteration the distance from Mahen remains the same and rotates in the counterclockwise direction by the angle z forms with the positive x-axis.

11. $|a| = \sqrt{3^2 + 4^2} = 5$, $|b| = \sqrt{5^2 + 12^2} = 13$
$ab = (3 + 4i)(5 + 12i) = -33 + 56i$
$|ab| = \sqrt{(-33)^2 + 56^2} = 65$
Hey! Notice that $5 \cdot 13 = 65$
Whoa!

12. **a.** $|w| = \sqrt{2^2 + 1^2} = \sqrt{5}$
b. Well, we saw in the previous problem that if we multiplied two complex numbers together, the resulting magnitude was the product of the magnitudes. We want a number with magnitude 5 and w has magnitude $\sqrt{5}$, so, hopefully, w^2 will have magnitude $\sqrt{5} \cdot \sqrt{5} = 5$.
Here goes nothin'!
$w \cdot w = 3 + 4i$ and $|ww| = \sqrt{3^2 + 4^2} = 5$
Yes!

13. Let's use the ideas from the previous problems. We want to square each complex number to ensure that our magnitude is a perfect square . . .

a. Note that $|4+i| = \sqrt{17}$
$(4+i)^2 = 15+8i$
$|(4+i)^2| = \sqrt{15^2+8^2} = \sqrt{17^2} = 17$
PPT: 8, 15, 17

b. Note that $|8+3i| = \sqrt{73}$
$(8+3i)^2 = 55+48i$
$|(8+3i)^2| = \sqrt{55^2+48^2} = \sqrt{73^2} = 73$
PPT: 48, 55, 73

c. Note that $|15+4i| = \sqrt{241}$
$(15+4i)^2 = 209+120i$
$|(15+4i)^2| = \sqrt{209^2+120^2} = \sqrt{241^2} = 241$
PPT: 120, 209, 241

d. Note that $|16+7i| = \sqrt{305}$
$(16+7i)^2 = 207+224i$
$|(16+7i)^2| = \sqrt{207^2+224^2} = \sqrt{305^2} = 305$
PPT: 207, 224, 305

e. Note that $|23+2i| = \sqrt{533}$
$(23+2i)^2 = 525+92i$
$|(23+2i)^2| = \sqrt{525^2+92^2} = \sqrt{533^2} = 533$
PPT: 92, 525, 533

f. Note that $|42+9i| = \sqrt{1845}$
$(42+9i)^2 = 1683+756i$
$|(42+9i)^2| = \sqrt{1683^2+756^2} = \sqrt{1845^2} = 1845$
PPT: 756, 1683, 1845

14. **a.** To find positive integers a, b, c, so that $abc = 72$ and $a+b+c = 14$, one way is to check all the factorizations abc of 72 to see which ones sum to 14. This is a fairly small number of cases. Since $72 = 2^3 3^2$, there are 12 divisors of 72, including the extreme cases of 1 and 72. However, the only divisors less than 14 are these: 1, 2, 3, 4, 6, 8, 9, 12. We can organize the search: pick each possible value of a, then check the sum for all possible values of b and c, avoiding duplicates by checking only $a \leqslant b \leqslant c \leqslant 12$. The results are organized in the table. We see the numbers $2, 6, 6$ and $3, 3, 8$ are the desired answer.

a	b	c	$a+b+c$
2	2	18	22
2	3	12	17
2	4	9	15
2	6	6	14
3	3	8	14
3	4	6	13

b. For a rectangular solid box, the surface area is the sum of the areas of the six rectangular sides; since opposite sites are congruent, this is the same as twice the sum of 3 adjacent rectangular areas. If the numbers a, b, c from (a) are the areas of these 3 rectangles, then the volume of the solid is $\sqrt{abc} = \sqrt{72}$. Thus two boxes, one with side areas 2, 6, and 6, and the other with areas 3, 3, 8, have the same surface area = 28 and the same volume $\sqrt{72}$.

What are the dimensions of such boxes? We can solve for the lengths of the edges, but the edges will not be integers. But once we have these numbers, we can scale all lengths by the same factor and get two boxes with integer side lengths and the same surface area and volume.

To solve for sides x, y, z, we solve the equations

$$\begin{aligned} yz &= a \\ xz &= b \\ xy &= c \end{aligned}$$

When $a = 2, b = 6, c, = 6$, the last two equations imply $z = y$ and the first equation says $z = y = \sqrt{2}$. Then $x = 3\sqrt{2}$ follows from the second equation. The sides of the first box are

$$(x, y, z) = (3\sqrt{2}, \sqrt{2}, \sqrt{2})$$

When $a = 3, b = 3, c, = 8$, the first two equations imply $x = y$ and the last equation says $x = y = 2\sqrt{2}$. Then $z = \frac{3}{4}\sqrt{2}$ follows from the second equation. The sides of the second box are

$$(x, y, z) = (2\sqrt{2}, 2\sqrt{2}, \frac{3}{4}\sqrt{2})$$

We can check that the surface area of each box is 28 and the volume is $\sqrt{72} = 6\sqrt{2}$.

To get two boxes with equal surface areas and equal volumes, but with integer sides, we can multiply all the lengths by $2\sqrt{2}$ to get new boxes with sides equal to $(12, 4, 4)$ and $(8, 8, 3)$, with surface area 224 and volume 192.

15. Let $w = 3 + 4i$. Then $|w|^2 = 25$ gives the Pythagorean triple 3, 4, 5. Scaling up by 25, this produces the non-primitive triple 75, 100, 125.

The square $w^2 = -7 + 24i$. The equation $|w^2|^2 = |w|^4 = 25^2$ produces the Pythagorean triple 7, 24, 25. This can be scaled up by 5 to give the triple 35, 120, 125.

The cube $w^3 = -117 + 44i$. The equation $|w^3|^2 = |w|^6 = 125^2$ produces the primitive Pythagorean triple 44, 117, 125.

These are all the Pythagorean triples with hypotenuse 125. The completeness can be checked by verifying a finite number of cases, or by reasoning further about factorization in the Gaussian integers.

16. Since $1105 = 5 \times 13 \times 17$, we can use relationships from early problems to state that $|3 + 4i|^2 = 5^2$, $|12 + 5i|^2 = 13$, and $|15 + 8i|^2 = 17^2$.

If we set $u = 3 + 4i$, $v = 12 + 5i$, and $w = 15 + 8i$, then $|uvw|^2 = 1105^2$.

It is also true that $|u\overline{v}w|^2 = 1105^2$, $|uv\overline{w}|^2 = 1105^2$, and $|u\overline{v}\overline{w}|^2 = 1105^2$. Each of these four equations gives a different primitive Pythagorean triple. There are four other products of u, v, w and their conjugates, but these products are the conjugates of the four products already given, so they produce the same triples.

For example $uvw = -943 + 576i$. Then $943^2 + 576^2 = 1221025 = 1105^2$. So 946, 576, 1105 is a primitive Pythagorean triple.

Using the other products $u\overline{v}w = 1073 + 264i$, $uv\overline{w} = -47 + 1104i$, and $u\overline{v}\overline{w} = 817 - 744i$, we obtain the three other primitive triples.

Problem Set 8

Opener

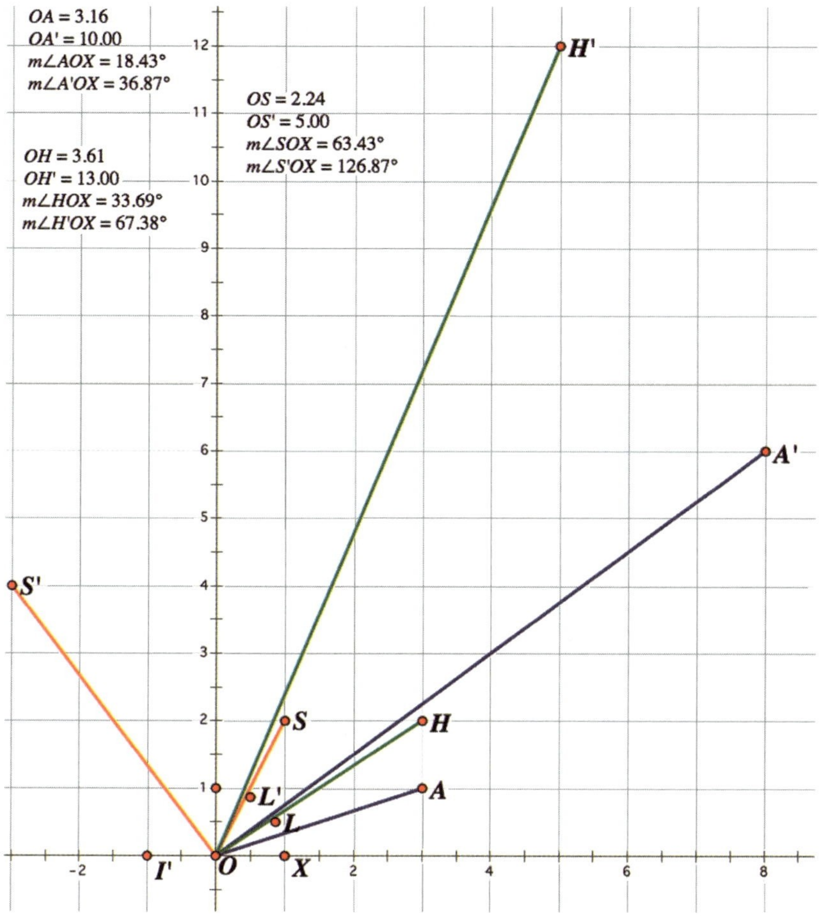

Hmmm . . . Well, it looks like the length gets squared. What is happening to the angles? Why aren't they getting squared too? Hey, they got doubled! What's up with that? I wonder what would happen if we cubed the complex number? Would we get a similar pattern? Hmm . . .

1. **a.** $z^0 = 1$
b. $z^1 = \frac{12}{13} + \frac{5}{13}i$
c. $z^2 = \frac{119}{169} + \frac{120}{169}i$
d. $z^3 = \frac{828}{2197} + \frac{2035}{2197}i$
e. $z^4 = -\frac{239}{28561} + \frac{28560}{28561}i$
f. $z^5 = -\frac{145668}{371293} + \frac{341525}{371293}i$

See the graph below for the plotted points (note that each tic mark is 1 unit):

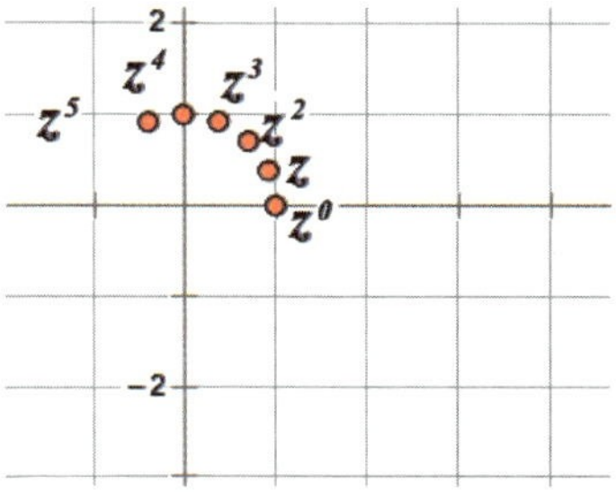

2. The angle formed each time is approximately 22.62°

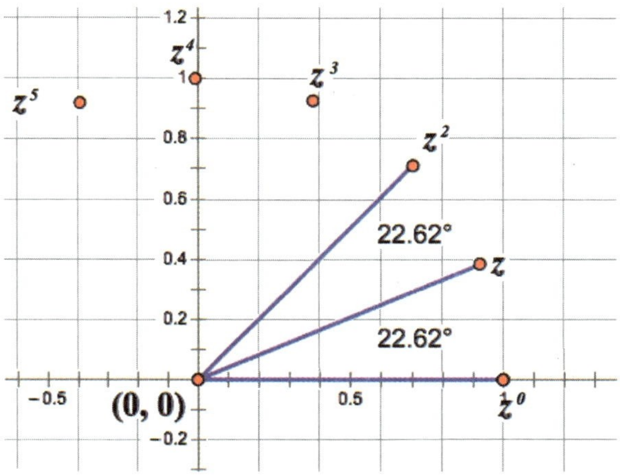

3. All numbers in the graph below have initial point at the origin and terminate at their points.
Adding numbers in the complex plan works the same as adding ordered pairs in the xy-plane.
Consider the numbers w and z in the graph below. If you were to take the segment for z and place it where w terminates, maintaining the direction of z, you will end up with $w + z$. In other words, imagine you are at the point w and you begin walking in the direction of z for the length of z. You would stop at $w + z$.

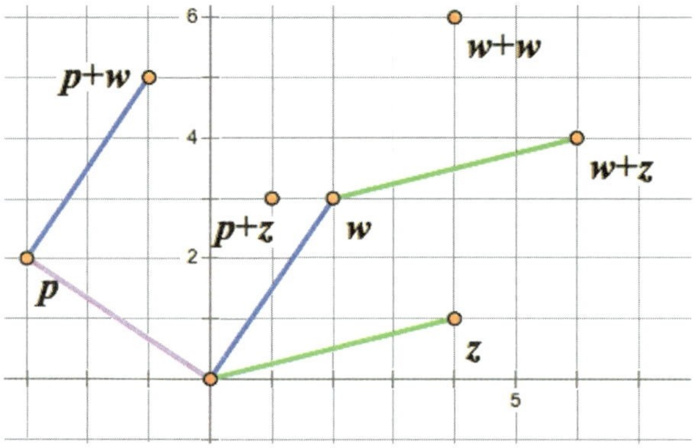

4. In the graph of the previous problem, you can see where $w + w = 2w$ falls, the result is a number with magnitude double the magnitude of w. When multiplying a complex number by a scalar, c, the magnitude gets multiplied by $|c|$.
The direction stays the same when multiplied by a positive scalar. When the scalar is negative, the direction is opposite (i.e. a 180° rotation from current direction).

5. If a complex number $a + bi$ is multiplied by i, we get $ai + b^2 = -b + ai$. The magnitude of $-b + ai$ is $b^2 + a^2$. So the magnitude of a complex number remains the same when it is multiplied by i.
If you imagine the complex number $a + bi$ as a vector from the origin to the point (a, b), $\langle a, b\rangle$, then the new vector resulting from multiplying by i is $\langle -b, a\rangle$. The direction of this vector is perpendicular to the direction of the original vector. In other words, the vector is rotated 90° in the counterclockwise direction.
You can also think of this in terms of slopes of lines:
$\frac{b}{a}$ vs. $-\frac{a}{b}$.

6. $wz = (1+i)(\sqrt{3}+i) = (\sqrt{3}-1) + i(\sqrt{3}+1)$,

$$\begin{aligned}|wz| &= \sqrt{(\sqrt{3}-1)^2 + (\sqrt{3}+1)^2} \\ &= \sqrt{3 - 2\sqrt{3} + 1 + 3 + 2\sqrt{3} + 1} \\ &= \sqrt{8} = 2\sqrt{2}\end{aligned}$$

Notice that $|w| = \sqrt{2}$ and $|z| = 2$. We could also find $|wz|$ by multiplying $|w||z|$. Thinking of $|wz|$ as a vector from the origin. The direction would be given by $\frac{1}{2\sqrt{2}}\langle\sqrt{3}-1, \sqrt{3}+1\rangle$.

7. When multiplying two complex numbers, the magnitude of the result is the product of the magnitudes of the two numbers. In other words, say z and w are two complex numbers. Then, $|zw| = |z| \cdot |w|$.
Additionally, to determine direction, it is useful to think of the angles formed. Let θ be the angle z forms with z^0 (i.e. θ is the angle z forms with the positive horizontal axis) and let ϕ be the angle for w. Then the angle for the product zw is $\theta + \phi$.

8. **a.** $|3+2i| = \sqrt{3^2+2^2} = \sqrt{13}$
b. $(3+2i)^2 = 5 + 12i$
$|(3+2i)^2| = \sqrt{25+144} = \sqrt{13^2} = 13$
c. The Pythagorean Triple is: 5, 12, 13

9. Pick your favorite $a + bi$, square it, and find the magnitude of this result! Just be careful of negative values that may result (See solutions from Day 6.)
I choose $2 + i$. $(2+i)^2 = 3 + 4i$, which has magnitude 5. Thus, the PT is 3, 4, 5.
How about $13 + 4i$. $(13+4i)^2 = 153 + 104i$, which has magnitude 185. Thus the PT is 104, 153, 185. Keep doing it!

10. **a.** 1,6,6 and 2,2,9
b. $2 \times 9 \times 9$ and $18 \times 3 \times 3$
c. $(x-a)(x-b)(x-c) = x^3 - x^2(a+b+c) + x(ab+ac+bc) - abc$
Notice that if a, b, c were dimensions of a box, $\frac{1}{2}$ the surface area would be the coefficient of x and the volume would be the constant.
When looking at the 3 numbers, as in part (a), the sum is the coefficient of x and their product is the constant.

11. **a.** Point Q has coordinates $(3, 0)$ and P has coordinates $(x, 0)$. Then $f(P)$ is the distance squared from P to Q, which is $(x-3)^2 + 0^2 = (x-3)^2 = x^2 - 6x + 9$. The graph of this quadratic function is a parabola (in the (x, z) plane!) which has its vertex (and its minimum) at $(3, 0)$. The function has the value 0 at this point.

b. Let point R have coordinates $(17, 0)$. The redefined function $f(P)$ is $(x-3)^2 + (x-17)^2 = 2x^2 - 40x + 298 = 2(x^2 - 20x + 149)$. The graph of this quadratic function is a parabola which has its vertex (and its minimum) at $(10, 0)$. The function has the value 98 at this point.

c. Let point S have coordinates $(1, 0)$. The redefined function $f(P)$ is $(x-3)^2 + (x-17)^2 + (x-1)^2 = 3x^2 - 42x + 299 = 3(x^2 - 14x + \frac{149}{3})$. The graph of this quadratic function is a parabola which has its vertex (and its minimum) at $(7, 0)$. The function has the value 152 at this point.

d. In each case the function is a parabola with leading coefficient equal to the number of points involved in the definition of f. The vertex is the mean or average of the points in the definition. For example, in the preceding case, the x-coordinate of the vertex is $\frac{3+17+1}{3}$.

12. **a.** Distance of point $P = (-1, 2)$ to point $(-1, 4)$ is

$$\sqrt{(-1+1)^2 + (4-2)^2} = \sqrt{2^2} = 2.$$

The distance from P to the line with equation $y = 1$ is $\sqrt{(1-2)^2} = \sqrt{1^2} = 1$. Yes; $2 = 2 \times 1$.

b. Distance of point $P = (-1, -2)$ to point $(-1, 4)$ is

$$\sqrt{(-1+1)^2 + (4+2)^2} = \sqrt{6^2} = 6.$$

The distance from P to the line with equation $y = 1$ is $\sqrt{(1+2)^2} = \sqrt{3^2} = 3$. Yes; $6 = 2 \times 3$.

c. Distance of point $P = (23, 14)$ to point $(-1, 4)$ is

$$\sqrt{(-1-23)^2 + (4-14)^2}$$
$$= \sqrt{24^2 + 10^2} = \sqrt{676} = 26.$$

The distance from P to the line with equation $y = 1$ is $\sqrt{(1-14)^2} = \sqrt{13^2} = 13$. Yes; $26 = 2 \times 13$.

d. Distance of point $P = (23, -14)$ to point $(-1, 4)$ is

$$\sqrt{(-1-23)^2 + (4+14)^2}$$
$$= \sqrt{24^2 + 18^2} = \sqrt{900} = 30.$$

The distance from P to the line with equation $y = 1$ is $\sqrt{(1+14)^2} = \sqrt{15^2} = 15$. Yes; $30 = 2 \times 15$.

e. Distance of point $P = (-91, -52)$ to point $(-1, 4)$ is

$$\sqrt{(-1+91)^2 + (4+52)^2}$$
$$= \sqrt{90^2 + 56^2} = \sqrt{11236} = 106.$$

The distance from P to the line with equation $y = 1$ is

$$\sqrt{(1+52)^2} = \sqrt{53^2} = 53.$$

Yes; $106 = 2 \times 53$.

f. The distance of point $P = (x, y)$ to point $(-1, 4)$ is $\sqrt{(-1-x)^2 + (4-y)^2}$.
The distance from P to the line with equation $y = 1$ is $\sqrt{(1-y)^2}$.
The distance to the point is twice the distance to the line is $\sqrt{(-1-x)^2 + (4-y)^2} = 2\sqrt{(1-y)^2}$. The relationship remains true if we square both sides, giving $(x+1)^2 + y^2 - 8y + 16 = 4(y^2 - 2y + 1) = 4y^2 - 8y + 4$.
This equation can be simplified to

$$3y^2 - (x+1)^2 = 12$$

Any point (x, y) is twice as far from $(-1, 4)$ as from the line $y = 1$ if and only if it satisfies this equation. This equation looks familiar ; it is the equation of a hyperbola.

13. The equation for points (x, y) three times as far from $(-1, 4)$ as from the line $y = 1$ is derived in the same way. The distance expressions are the same.

The distance to the point is twice the distance to the line is $\sqrt{(-1-x)^2 + (4-y)^2} = 3\sqrt{(1-y)^2}$. The relationship remains true if we square both sides, giving $(x+1)^2+y^2-8y+16 = 9(y^2-2y+1) = 9y^2 - 18y + 9$.

This equation can be rearranged as $(x + 1)^2 = 9y^2 - y^2 - 18y + 8y + 9 - 16$, or

$$(x + 1)^2 - 8y^2 + 10y + 7 = 0$$

This set of points is shown in the graph below.

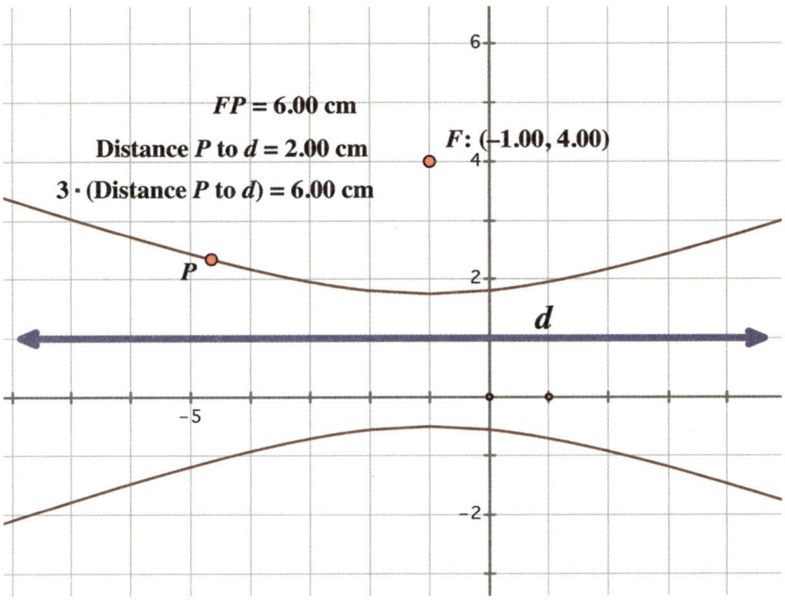

14. The equation for points (x, y) one-half times as far from $(-1, 4)$ as from the line $y = 1$ is derived in the same way. The distance expressions are the same.

The distance to the point is twice the distance to the line is $2\sqrt{(-1-x)^2 + (4-y)^2} = \sqrt{(1-y)^2}$. The relationship remains true if we square both sides, giving $4(x + 1)^2 + 4(y - 4)^2 = (y - 1)^2$.

Expanding this gives $4(x + 1)^2 + 4y^2 - 32y + 64 - y^2 + 2y - 1$ and combining gives

$$4(x + 1)^2 + 3y^2 - 30y + 63 = 0$$

15. The triangle is a right triangle, since the sides are a Pythagorean triple. The perimeter equals $8 + 15 + 17 = 40$ and the area equals $\frac{8 \times 15}{2} = 60$. If the length of each side is multiplied by nonzero k, then the perimeter becomes $40k$ and the area becomes $60k^2$. Equating the two numbers, we get $40k = 60k^2$, which is true for $k = 2/3$.

The sides of the similar triangle are $16/3$, 10, $34/3$. The perimeter and area are $80/3$.

16. The easiest triangles to check are the isosceles triangles. Each such triangle can be dissected into two congruent right triangle "halves", where the sides of the isosceles triangle are c, c, $2a$ with c the hypotenuse of a right triangle and a one of the legs. The triangle has a given perimeter $2s$ if $a+c = s$. For such a triangle, the area of the isosceles triangle is $a\sqrt{c^2 - a^2}$, twice the area of a right triangle.

One practical way to search for two non-congruent isosceles triangles with the same perimeter and the same area is to fix a value of s and then compute and compare the areas of the right triangles for all possible a values, with $c = s - a$ and $a < s/2$ so that $a < c$. For very small s there are not enough cases to make success likely. But starting with $s = 10$, one soon comes to $s = 14$ and finds two equal areas.

The right triangle with $a = 3$ and $c = 11$ and the right triangle with $a = 6$ and $c = 8$ both have area $6\sqrt{7}$. Therefore, the isosceles triangles with sides 11, 11, 6 and 8, 8, 12 both have perimeter 28 and area $12\sqrt{7}$.

17. Making numerical experiments with geometry software, the radius of the inscribed circle in a right triangle with leg lengths a and b and hypotenuse c appears to be

$$\frac{ab}{a+b+c} = \frac{ab}{a+b+\sqrt{a^2+b^2}}$$

The reasoning behind this formula will be taken up in the Important Stuff in the next problem set.

18. To see the effect of multiplying a complex number w by $z = a + bi$, one can compare these two figures. The first figure shows that the complex numbers 0, a, $a + ib$ and ib are the vertices of a rectangle, since the number $ib = bi$ is perpendicular to a because it is on the y-axis. The magnitude $|z|$ of $a + ib$ is the length of the hypotenuse of the rectangle. And the direction of $a + ib$ is the oriented angle that the diagonal makes with the x-axis.

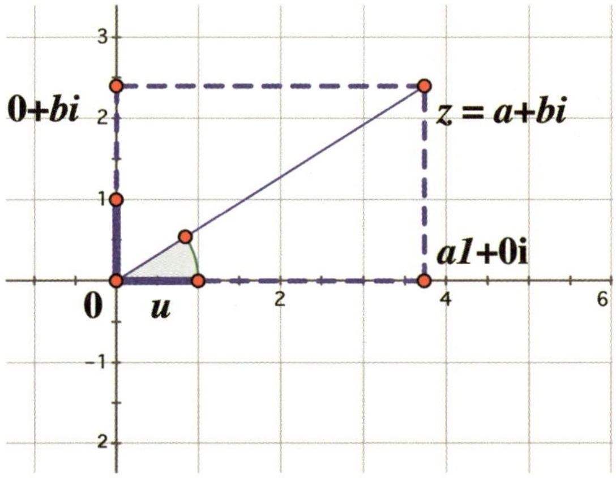

The second figure shows a complex number w and the product of w with the numbers 0, a, $z = a + ib$ and ib. in effect this picture is the same figure, scaled and rotated. The role of the unit $1 + i0$ in the first figure as been replaced by w in the new figure.

To give more detail, we have seen that aw is a complex number in the same direction as w (or opposite, if $a < 0$) with length or magnitude $|aw| = |a||w|$. We have also seen that iw is the complex number obtained by rotating w about the origin by 90 degrees, with $|iw| = |w|$. Then biw is also perpendicular to w with $|ibw| = |b||w|$.

By the parallelogram rule for adding vectors, the sum of these two complex numbers $aw + biw = (a + bi)w = zw$ is the fourth vertex of the parallelogram formed by 0, aw and ibw. Since aw and ibw are perpendicular, the parallelogram is a rectangle. Since the lengths in the original figure are all multiplied by $|w|$, the figures are similar. In particular the right triangles formed by the sides of the rectangles and the diagonals are similar, so that $|zw| = |z||w|$.

Also, since the triangles are similar, the oriented angle that the diagonal makes with the base of the rectangle is the same. This means the the direction angle of zw is the direction angle of z added to the direction angle of w. One caveat: this figure shows the simplest case when both complex numbers are in the first quadrant and the angles are positive. To see this in general, one must also consider the other quadrants, where the angles may be negative or greater than a straight angle. One can show the addition relation holds in all cases.

Another way of expressing this angle relation is that the direction of w is rotated by the angle of rotation that rotates the positive s-axis to the direction of z.

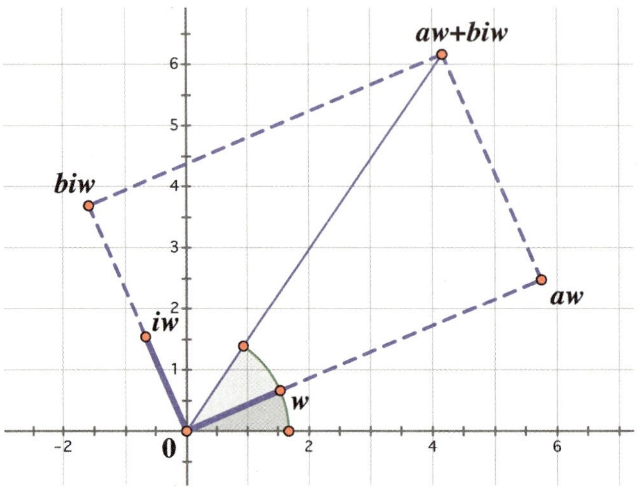

19. To find a primitive Pythagorean triple whose hypotenuse has length 13^3, we the method of taking the argument square of a complex power.

$$|5 + 12i|^2 = 5^2 + 12^2 = 13^2$$

Squaring this, $(5 + 12i)^2 = -119 + 120i$, so

$$|-119 + 120i|^2 = 119^2 + 120^2 = 13^4$$

Taking the third power, $(5+12i)^3 = 5+12i)(-119+120i) = -2035 - 828i$, so

$$|2035 - 828i|^2 = 2035^2 + 828^2 = 13^6 = (13^3)^2$$

So the Pythagorean triple is 828, 2025, 13^3.

20. Let $z = m + ni$. Then $z^2 = (m^2 - n^2) + 2mni$. We have seen in Problem 5 and Problem 18 that $|z^2| = |zz| = |z||z| = |z|^2$. Writing this out in detail, we see

$$(m^2 - n^2)^2 + (2mn)^2 = |z^2| = |z|^2 = (m^2 + n^2)^2$$

21. Given that a is a complex number with $|a| = 5$ and b is a complex number with $|b| = 13$, what can the magnitude of $a + b$ be?

Consider the triangle with vertices 0, a, and $a + b$. Then the lengths of the sides are $|a| = 5$, $|b| = 13$, and $|a + b|$. By the triangle inequality, the length of any side of a triangle is less than the sum of the other two sides: $|a+b| < |a|+|b|$. There is a special case when 0, a, and b are collinear, i.e., when one of the complex numbers is a real multiple of the other. In this case $|a + b| = |a| + |b|$ if the real multiple is positive and $|a + b| = ||a| - |b||$ if the multiple is negative real. So in all the cases, $|a + b| \leqslant |a| + |b| = 5 + 13 = 18$.

22. Suppose that $x = a^2+b^2$ and $y = c^2+d^2$. The same equations using complex numbers are $x = |a+bi|^2$ and $y = |c + di|^2$.

Then $xy = |a + bi|^2|c + di|^2 = |(a + bi)(c + di)|^2 = |(ac - bd) + (ad + bc)i|^2$.

This implies that $xy = (a^2 + b^2)(c^2 + d^2) = (ac-bd)^2+(ad+bc)^2$. This can be checked directly by expanding the expressions.

23. In a regular n-gon, the central angle formed by two adjacent vertices with vertex at the origin has measure $360/n$ degrees.

a. If $(-1, 0)$ is one of the vertices, this means that $180 = 360/2 = 360k/n$ for some integer k. This means that $n = 2k$, so n must be even. Conversely, if $n = 2k$ and $k = n/2$, then $(-1, 0)$ is the k-th vertex from $(1, 0)$, moving in a counterclockwise direction.

b. If $(0, -1)$ is one of the vertices, this means that $90 = 360/4 = 360k/n$ for some integer k. This means that $n = 4k$, so n is divisible by 4. Conversely, if $n = 4k$ and $k = n/4$, then $(0, -1)$ is the k-th vertex from $(1, 0)$ moving in a clockwise direction.

c. Let the vertices of the 12-gon be labeled by the direction angle of the vertex, so the vertices are labeled $30k$, with $k = 0, 1, \ldots, 11$.

The vertices of the equilateral triangle are the vertices with $k = 0, 4, 8$, the numbers that are the multiples of $12/3$.

The vertices of the square are the vertices with $k = 0, 3, 6, 9$, the numbers that are the multiples of $12/4$.

The vertices of the regular hexagon are the vertices with $k = 0, 2, 4, 6, 8, 10$, the numbers that are the multiples of $12/6$.

We conclude that the vertices of the dodecagon that are *not* vertices of the smaller n-gons are the vertices with $k = 1, 5, 7, 11$.

Problem Set 9

Opener

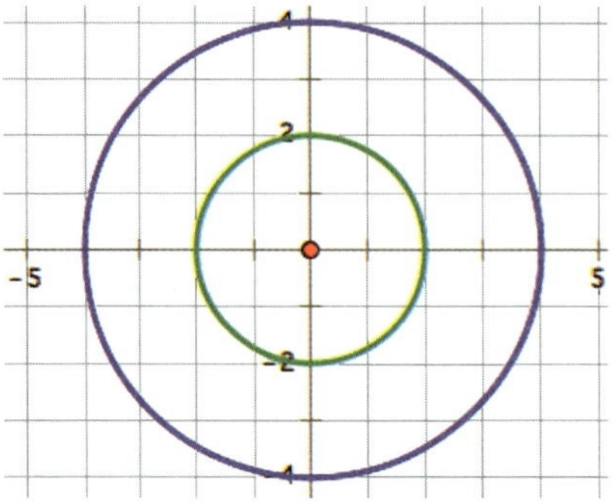

Let's pick something simple. How about $v = 2i$. $v^2 = -4$. Complex numbers with magnitude 2 will be numbers $v = a + bi$ where $a^2 + b^2 = 4$. So all points on a circle with radius 2 centered at the origin.

Now, if we square these numbers, we will have a magnitude of 4. So if $v^2 = c + di$, then $c^2 + d^2 = 16$. The squares lie on the circle with radius 4 centered at the origin.
Circle takes the square! Never mind . . .

BUT! If we square all of our points on the radius 2 circle, will the results generate all the points on the circle with radius 4? We know from the previous problem set that the angles change, so a point in a particular place on the smaller circle will not necessarily be in the corresponding place on the larger circle. So how do we get all the points on the circle?

1. Square first, then find the magnitude . . . just like before! This is old hat for us (see Problem Set 7)! Just be sure they are Primitive PTs!
Here are some ridiculous ones:
$(99 + 14i)^2 = 9605 + 2772i$, which has magnitude 9997, so the PPT is 2772, 9605, 9997.
$(45 + 8i)^2 = 1961 + 720i$, which has magnitude 2089, so the PPT is 720, 1961, 2089.
$(86 + 51i)^2 = 4795 + 8772i$, which has magnitude 9997, so the PPT is 4795, 8772, 9997.
$(101+100i)^2 = 201+20200i$, which has magnitude 20201, so the PPT is 201, 20200, 20201.
$(400+307i)^2 = 65751+245600i$, which has magnitude 254249, so the PPT is 65751, 245600, 254249.

2. For the height, drop a perpendicular from the top of the triangle to the base where the length is 25. This breaks up the base into two segments: $25 - x$ and x
$x^2 + h^2 = 17^2$ and $(25 - x)^2 + h^2 = 28^2$
$h^2 = 17^2 - x^2$, so we can rewrite the 2nd equation:
$(25 - x)^2 + 17^2 - x^2 = 28^2$
$x = \frac{25^2+17^2-28^2}{50} = \frac{13}{5} \Rightarrow h = \frac{84}{5}$
$A = \frac{1}{2} \cdot 25 \cdot \frac{84}{5} = 210$

3. Consider the picture of a circle inscribed inside a triangle below:

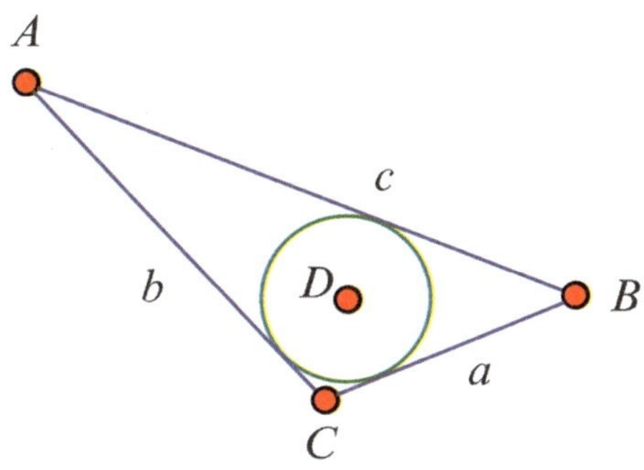

The large triangle, $\triangle ABC$ is comprised of three smaller triangles: $\triangle BCD$, $\triangle CAD$, $\triangle ABD$. These three triangles all have height r, which corresponds to the radius of the inscribed circle. Thus the area of $\triangle ABC$ is $A = \frac{1}{2}ar + \frac{1}{2}br + \frac{1}{2}cr = \frac{1}{2}r(a + b + c) = \frac{1}{2}rP$

4. Let's use $A = \frac{1}{2}Pr$.
$P = a + b + c = 25 + 28 + 17 = 70$ and $A = 210$.
Plug in values: $210 = \frac{1}{2}(70)r \Rightarrow r = 6$

5. In the figure, let the height be h and the lengths of the two segments on the base be $c - x$ and x. So the Pythagorean Theorem provides two equations: $x^2 + h^2 = b^2$ and $(c - x)^2 + h^2 = a^2$.

Substituting $h^2 = b^2 - x^2$ from the first equation into the second gives $c^2 - 2cx + x^2 + b^2 - x^2 = a^2$.

This equation says $2cx = b^2 + c^2 - a^2$, so $x = \frac{1}{2c}(b^2 + c^2 - a^2)$.

We can substitute x^2 into the first equation to solve for h. Since

$$\begin{aligned} x^2 &= \frac{1}{4c^2}(b^2 + c^2 - a^2)^2 \\ &= \frac{1}{4c^2}((b^2 + c^2)^2 - 2a^2(b^2 + c^2) + a^4) \end{aligned}$$

then

$$\begin{aligned} h &= \sqrt{b^2 - x^2} \\ &= \frac{1}{2c}\sqrt{4b^2c^2 - ((b^4 + 2b^2c^2 + c^4) - (2a^2b^2 + 2a^2c^2) + a^4)} \end{aligned}$$

Collecting like terms and grouping, we get

$$h = \frac{1}{2c}\sqrt{2(b^2c^2 + a^2c^2 + a^2b^2) - (a^4 + b^4 + c^4)}$$

Since the triangle area $A = ch/2$,

$$A = \frac{1}{4}\sqrt{2(b^2c^2 + a^2c^2 + a^2b^2) - (a^4 + b^4 + c^4)}$$

6. Also, if r is the radius of the inscribed circle, the area $A = \frac{1}{2}r(a + b + c)$. So

$$\begin{aligned} r &= \frac{2A}{a + b + c} \\ &= \frac{\sqrt{2(b^2c^2 + a^2c^2 + a^2b^2) - (a^4 + b^4 + c^4)}}{2(a + b + c)} \end{aligned}$$

7. Note that both sets of numbers are PT's, so both triangles are right triangles. Let's scale the 8-15-17 triangle by a factor of 5 to get a 40-75-85 triangle. Scale the 5-12-13 triangle by a factor of 8 to get a 40-96-104 triangle. Now they both have a side with the same length. Put them together:

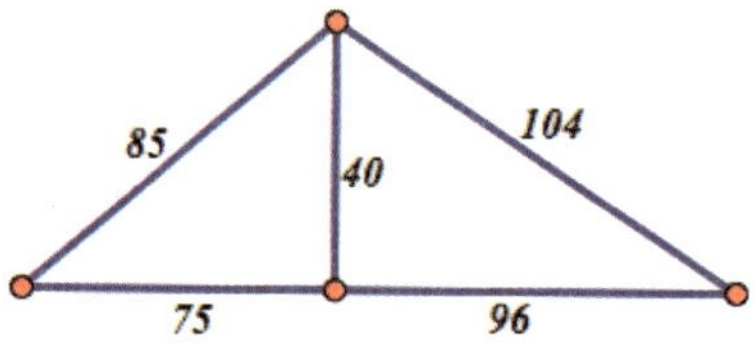

8. Ah, more with Heronian triangles. We know that both sets of numbers are PT's, so both triangles are right triangles. Let's scale the 8-15-17 triangle by a factor of 4 to get a 32-60-68 triangle. Scale the 5-12-13 triangle by a factor of 5 to get a 25-60-65 triangle. Now they both have a side with the same length. Put them together:

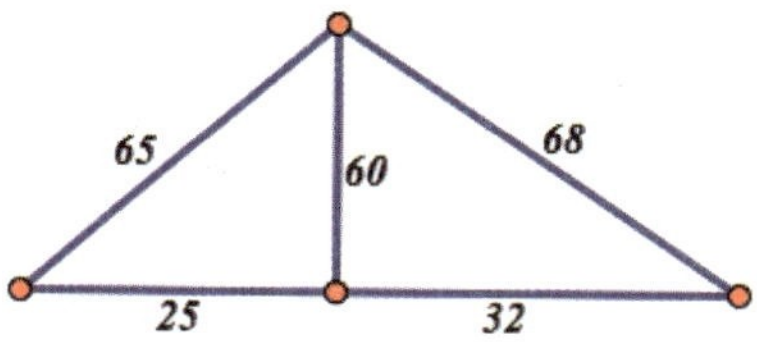

Now keep the 8-15-17 triangle as-is and scale the 5-12-13 triangle by a factor of 3 to get a 15-36-39 triangle. Slap 'em together!

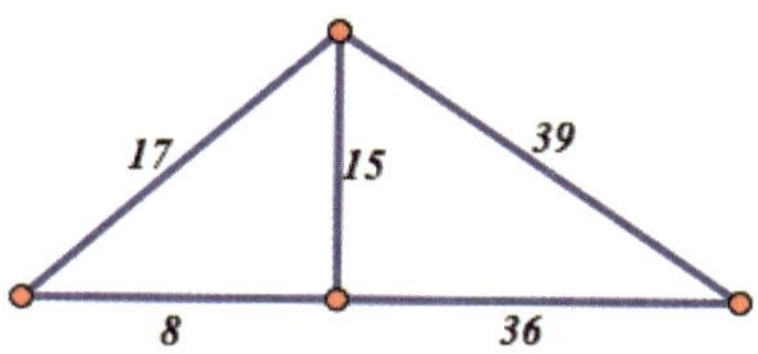

9. **Ridiculous Heronian Triangle 1:**

Using the PPTs: 2772-9605-9997 and 4795-8772-9997

Multiply the first set by 731 and the second set by 231 to get:

2026332 – 7021255 – 7307807 (call it $a - b - c$)

and 1107645 – 2026332 – 2309307 (call it $d - b - e$)

In order for the picture below to not get too cluttered with these crazy numbers, we will use the corresponding letters for labels.

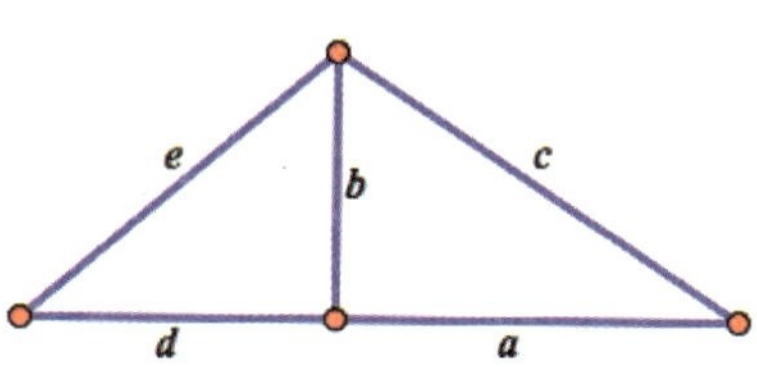

Ridiculous Heronian Triangle 1:

Using the PPTs: 65751 – 245600 – 254249 and 4795 – 8772 – 9997

Multiply the first set by 959 and the second set by 49120 to get:

63055209 – 235530400 – 243824791 (call it p – q – r) and 235530400 – 430880640 – 491052640 (call it q – s – t)

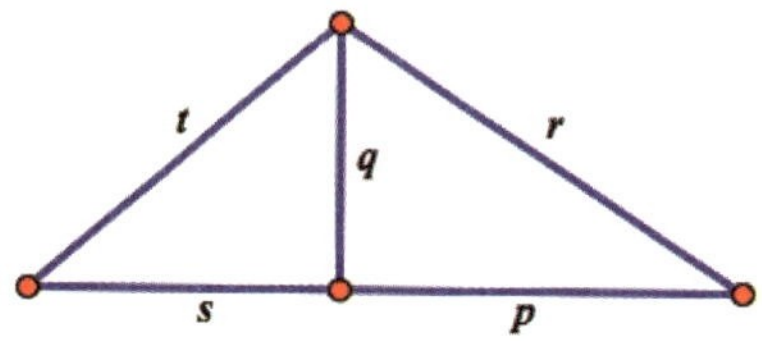

Well, you get the idea . . .

10. We discussed this in Problem Set 8. Imagine you are at the point w and you begin walking in the direction of z for the length of z. You would stop at $z + w$. Or you can start at the point z and begin walking in the direction of w for the length of w. You would stop at $w + z = z + w$.

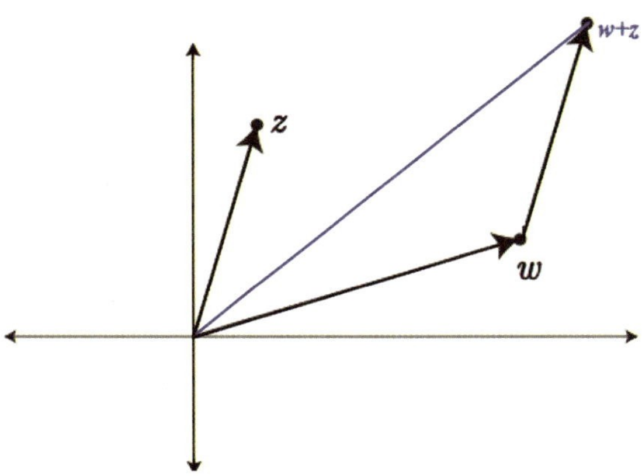

11. The product is $(1+2i)(1-2i) = (1^2+2^2)+(2-2)i = 5$.

12. $|1 + 2i| = \sqrt{1^2 + 2^2} = \sqrt{5}$.

13. **a.** The triangle with vertices $3 + i$, $3 + 2i$, i is a right triangle with one leg of length 3 parallel to the x-axis and the other leg of length 1.

b. The three products are $(1+2i)(3+i) = 1+7i$, $(1+2i)(3+2i) = -1+8i$, $(1+2i)(i) = -2+i$.

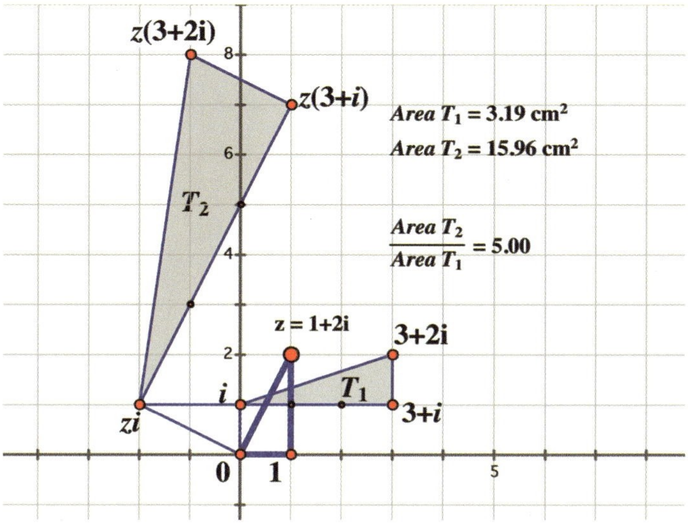

c. The new triangle is also a right triangle. The leg parallel to the x-axis has been transformed to a segment parallel to $1 + 2i$ with length equal to 3 times the magnitude of $1 + 2i$. The other leg is perpendicular and of length equal to the magnitude of $1 + 2i$.

d. The new triangle is similar to the old triangle with ratio of lengths $\sqrt{5}$. The ratio of the new area to the original area is 5.

14. If the sides of the box are 4, 10, 15, then the volume is $4 \times 10 \times 15 = 600$. The surface is made up of 3 pairs of congruent rectangles of area $4 \times 10 = 40$, $4 \times 15 = 60$, $10 \times 15 = 150$, so the total area is $2(40 + 60 + 150) = 500$.

If the sides of the box are 5, 6, 20, then the volume is $5 \times 6 \times 20 = 600$. The surface is made up of 3 pairs of congruent rectangles of area $5 \times 6 = 30$, $5 \times 20 = 100$, $6 \times 20 = 120$, so the total area is $2(30 + 100 + 120) = 500$.

15. **a.**

$$\begin{aligned}(x-4)(x-10)(x-15)\\ &= x^3 - (4+10+15)x^2\\ &\quad + (4\times 10 + 4\times 15 + 10\times 15)x\\ &\quad - 4\times 10\times 15\\ &= x^3 - 29x^2 + 250x - 600.\end{aligned}$$

b.

$$\begin{aligned}(x-4)(x-10)(x-15)\\ &= x^3 - (4+10+15)x^2\\ &\quad + (4\times 10 + 4\times 15 + 10\times 15)x\\ &\quad - 4\times 10\times 15\\ &= x^3 - 29x^2 + 250x - 600.\end{aligned}$$

c.

$$\begin{aligned}(4x-1)(10x-1)(15x-1)\\ &= 600x^3 - 250x^2 + 29x - 1.\end{aligned}$$

d.

$$\begin{aligned}(x-\frac{1}{4})(x-\frac{1}{10})(x-\frac{1}{15})\\ &= \frac{1}{4}\frac{1}{10}\frac{1}{15}((4x-1)(10x-1)(15x-1)\\ &= \frac{1}{600}(600x^3 - 250x^2 + 29x - 1)\end{aligned}$$

16. We see from the expansion of the products in the preceding problem that, for lengths a, b, c, the sum $a+b+c$, is the coefficient of x^2 in the polynomial $(x-a)(x-b)(x-c)$. The total edge length of the box is four times this sum. The volume of the box is the product abc, which is the constant term. Thus if we expand this for $a = \frac{1}{5}$, $b = \frac{1}{6}$, $c = \frac{1}{20}$, then we get the same coefficients of x^2 and the same constant term as in 15(d), so the total edge lengths and volume are the same as for $\frac{1}{4}$ by $\frac{1}{10}$ by $\frac{1}{15}$.

$$\begin{aligned}(x-\frac{1}{5})(x-\frac{1}{6})(x-\frac{1}{20})\\ &= \frac{1}{600}((5x-1)(6x-1)(20x-1)\\ &= \frac{1}{600}(600x^3 - 250x^2 + 31x - 1)\end{aligned}$$

17. The task here is to figure out an expression s so that this formula computes the area of a triangle with sides a, b, c.

$$A = \sqrt{s(s-a)(s-b)(s-c)}$$

We already have a formula for the area developed in Problem 5. We should be able to rearrange this formula to express it in this new form. However, it will be a lot easier to do this if we already know or can conjecture how s is expressed in terms of a, b, c. (Of course we may know an expression for s already, because this is a famous formula. But next we will show that the expression for s is really forced on us.)

What do we know about s? First, the formula for A must be symmetric in the variables a, b, c since the area does not depend on the order of the sides. So the expression s should be symmetric in the variables for A to be symmetric. Also, thinking of units and dimension, if the formula for A provides a quantity in square units, s should be in length units; the simplest expression of such a quantity is a first-degree polynomial expression in the variables: $s = Ka + Lb + Mc$. Since s is symmetric, all the constants should be the same and $s = K(a+b+c)$ for some number K.

Another thing that must be true about the expression for A is that it should be zero whenever $a+b-c = 0$ or $a+c-b = 0$ or $b+c-a = 0$, since for those cases the 3 vertices are collinear. This suggests that the expression $s-a = (K-1)a+Kb+Kc$ should be a multiple of $b+c-a$, which will be true if $K = \frac{1}{2}$.

Thus, if there is such a formula with s, the most likely and perhaps only possibility for s is to set $s = \frac{1}{2}(a+b+c)$, which is one-half the perimeter. In this case, $s-a = \frac{1}{2}(b+c-a)$, $s-b = \frac{1}{2}(a+c-b)$ and $s-c = \frac{1}{2}(a+b-c)$.

Then the new area formula with s takes this form:

$$A = \frac{1}{4}\sqrt{(a+b+c)(b+c-a)(a+c-b)(a+b-c)}$$

The formula from Problem 5 is of this form:

$$A = \frac{1}{4}\sqrt{2(b^2c^2+a^2c^2+a^2b^2)-(a^4+b^4+c^4)}$$

The two formulas will be equivalent provided that the squares are equal, so to check the equivalence, we need to show that

$$2(b^2c^2 + a^2c^2 + a^2b^2) - (a^4 + b^4 + c^4)$$
$$= (a + b + c)(b + c - a)(a + c - b)(a + b - c)$$

It is possible to multiply out the expression on the right and expand it to a sum of monomials and then compare with the expression on the left. The next equations do this step by step by some grouping of terms so that the expressions do not become too long at once.

$$\begin{aligned}
&(a + b + c)(b + c - a)(a + c - b)(a + b - c)\\
&\quad = ((a + b) + c)((a + b) - c)(c + (b - a)(c - (b - a))\\
&\quad = ((a + b)^2 - c^2)((c^2 - (b - a)^2)\\
&\quad = -(a + b)^2(a - b)^2 + c^2((a + b)^2 + (a - b)^2) - c^4\\
&\quad = -(a^2 - b^2)^2 + c^2(2a^2 + 2b^2) - c^4\\
&\quad = -a^4 + 2a^2b^2 - b^4 + 2c^2a^2 + 2c^2b^2 - c^4\\
&\quad = 2(b^2c^2 + a^2c^2 + a^2b^2) - (a^4 + b^4 + c^4)
\end{aligned}$$

18. The perimeter of a triangle with sides 17, 25, 28 is $17 + 25 + 28 = 70$, so the quantity $s = 70/2 = 35$ in the new area formula.

Then the area is

$$A = \sqrt{35 \times 18 \times 10 \times 7} = 210$$

If another triangle with the same perimeter and area, then the value of s is still 35. it is useful to give names to the quantities in the area formula. Set $x = s - a = 35 - a$, $y = s - b = 35 - b$, $z = s - c = 35 - c$. Then $x + y + z = 105 - (a + b + c) = 105 - 2s = 35$. The area $A = \sqrt{35xyz} = 210$.

So to find another triangle with the same perimeter and area, it is sufficient to find x, y, z so that $x + y + z = 35$ and $35xyz = 210^2$. The last equation can be rewritten as $xyz = 1260 = 5 \times 7 \times 2^2 \times 3^2$.

If the variables x, y, z are integers, then the possible solutions are obtained by allocating the factors of 1260 to the 3 variables in as many ways as possible. We see immediately that if 35 divides any one of the variables then the sum $x + y + z > 35$ so that possibility is eliminated. Therefore, the factor 7 and the factor 5 have to occur in different variables. Without loss of generality, we can set $x = 7x'$, $y = 5y'$ and then $x'y'z = 36$. It is not hard to check the sum $x + y + z = 35$ for the small number of cases and find that $x = 14$, $y = 15$, and

$z = 6$ gives the sum 35. Then solving for the original variables, we get $a = 21$, $b = 20$ and $c = 29$ are the sides of a triangle with the same perimeter and area as the original one with sides of 17, 25, 28.

19. The formula for area can be used find triangles with sides $a = 15$, $b = 7$, $c = x$ so that, numerically, the area of the triangle equals the perimeter.

It should be noted that not every value of x corresponds to the side of a triangle that exists, since the three sides have to satisfy the triangle inequality. In this case, this means that $x+7 > 15$, $x+15 > 7$ and $15+7 > x$, so $8 < x < 22$.

The perimeter equals $22 + x$ and $s = 11 + x/2$, and the quantities that appear in the area formula are

$$s - a = \frac{x}{2} - 4; \quad s - b = \frac{x}{2} + 4; \quad s - c = -\frac{x}{2} + 11.$$

Equating the perimeter with the area gives

$$2(11 + \frac{x}{2}) = \sqrt{(11 + \frac{x}{2})(11 - \frac{x}{2})(\frac{x}{2} + 4)(\frac{x}{2} - 4)}$$

Squaring both sides:

$$4(11 + \frac{x}{2})^2 = (11 + \frac{x}{2})(11 - \frac{x}{2})(\frac{x}{2} + 4)(\frac{x}{2} - 4)$$

The common factor $(11 + \frac{x}{2}))$ can be factored from both sides and terms can be collected onto on side of the equation.

$$-(11 - \frac{x}{2})(\frac{x}{2} + 4)(\frac{x}{2} - 4) + 4(11 + \frac{x}{2}) = 0$$

Also, for simplicity in writing we temporarily substitute the variable $y = x/2$ and expand the products to get

$$(y - 11)(y^2 - 16) + 4(11 + y) = y^3 - 11y^2 - 12y + 220$$

The graph of this cubic polynomial appears to have a root at $y = 10$. That this is true can be checked by substitution. This implies $c = x = 2y = 20$. There is one more thing to be checked: Do the three lengths $a = 15$, $b = 7$ and $c = 20$ satisfy the triangle inequality? We check that $15 + 7 = 22 > 20$ and conclude that a triangle with these sides does exist.

Thus one of the desired triangles has sides $a = 15$, $b = 7$ and $c = 20$.

Is there another triangle? The cubic polynomial can be factored by $y = 10$ to get $y^3 - 11y^2 - 12y + 220 = (y-10)(y^2 - y - 22)$. The other two roots of the polynomial can be found by applying the quadratic formula to $y^2 - y - 22$. The roots are

$$y = \frac{1+\sqrt{89}}{2} > 0; \quad y = \frac{1-\sqrt{89}}{2} < 0$$

The positive root provides another length $c = x = 2y = 1 + \sqrt{89}$ as the side of a triangle with perimeter equal to area. The value is approximately equal to 10.43. It does satisfy the constraints of the triangle inequality, so the triangle exists.

Thus a second one of the desired triangles has sides $a = 15$, $b = 7$ and $c = 1 + \sqrt{89}$.

20. The distance from point $P = (x, y)$ to point $(-1, 4)$ is $\sqrt{(x+1)^2 + (y-4)^2}$. The distance to the line $y = 1$ is $\sqrt{(y-1)^2} = |y - 1|$.

If the distance from P to $(-1, 4)$ is α times the distance to the line, then

$$\sqrt{(x+1)^2 + (y-4)^2} = \alpha\sqrt{(y-1)^2}$$

Squaring and moving all the terms to one side:

$$\begin{aligned}&(x+1)^2 + (y-4)^2 = \alpha^2(y-1)^2 \\ &\quad\Longleftrightarrow (x+1)^2 + (y-4)^2 - \alpha^2(y-1)^2 = 0\end{aligned}$$

Then expanding:

$$(x+1)^2 + (1-\alpha^2)y^2 + (2\alpha^2 - 8)y + (16 - \alpha^2) = 0$$

Exploring the set of these points with The Geometer's Sketchpad, it appears that the locus is a hyperbola if $\alpha > 1$ and an ellipse if $0 < \alpha < 1$ and likely a parabola when $\alpha = 1$. In the equation this corresponds to cases when the sign of the coefficient of y^2 is negative (thus different from the sign of the coefficient of x^2), the sign of the coefficient of y^2 is positive, or the coefficient of y^2 is equal to zero.

21. The complex number $z = a + bi = \cos\theta + i\sin\theta$.

a. Computing the square, $z^2 = (\cos^2\theta - \sin^2\theta) + i(2\cos\theta\sin\theta)$. But $|z^2| = 1^2 = 1$ and the direction of z^2 is 2θ, so also $z^2 = \cos 2\theta + i\sin 2\theta$. This gives formulas for $\cos 2\theta$ and $\sin 2\theta$.

b. Continuing with the same idea,

$$\begin{aligned} z^3 &= \cos^3\theta + i3\cos^2\theta\sin\theta \\ &\quad - 3\cos\theta\sin^2\theta - i\sin^3\theta \\ &= (\cos^3\theta - 3\cos\theta\sin^2\theta) \\ &\quad + i(3\cos^2\theta\sin\theta - \sin^3\theta) \\ &= \cos 3\theta + i\sin 3\theta \end{aligned}$$

Problem Set 10

Opener

If v has magnitude 2, then v^3 will have magnitude 8. All the complex numbers $v = a + bi$ will lie on the circle with radius 2, centered at the origin. As we saw in Day 9's opener, it makes sense to have v^3 lie on the circle with radius 8. For example, $w = -\sqrt{2} + \sqrt{2}i$ has magnitude 2 and $w^3 = 4\sqrt{2} + 4\sqrt{2}i$, which has magnitude 8. Similarly with $v = -2 + 0i$ . . .

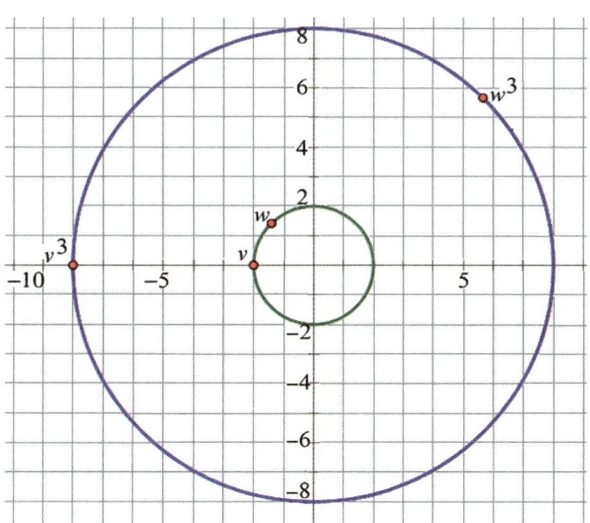

Using $v = -2 + 0i$, $v + i = -2 + i$. This moves v vertically upward by 1 unit.

$v + 3 = 1 + 0i$, which moves v horizontally by 3 units, since 3 is being added to the real part and the imaginary part remains unaffected.
$2iv = -4i$, which rotates v counterclockwise by $90°$ (Didn't we see this on Day 7?) I think the i is making this happen. Hmmm . . . Additionally, the magnitude is increased by a factor of 2.
$v^2 + v = 2$, which rotates v by $180°$. Does this hold for other points? Let's go ahead and check using our $w = -\sqrt{2} + \sqrt{2}i$ from earlier. $w^2 + w = -\sqrt{2} + i(\sqrt{2} - 4)$. $v^2 + v$ is 4 units from v. Hmmm . . . what is happening here? Perhaps it would help to think about v^2 first and then what happens when we add or subtract v.

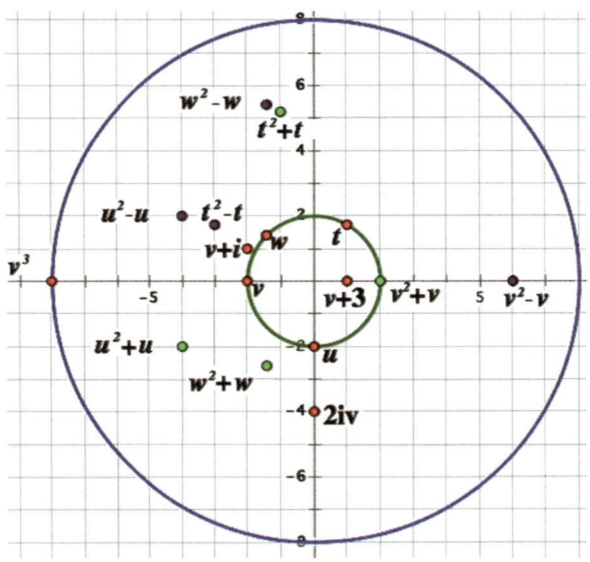

1. **Heronian Triangle 1:**

   Scale the 5-12-13 triangle by a factor of 17 to get the 85-204-221 triangle. Leave the 85-132-157 triangle as is and put them together as in the figure below:

   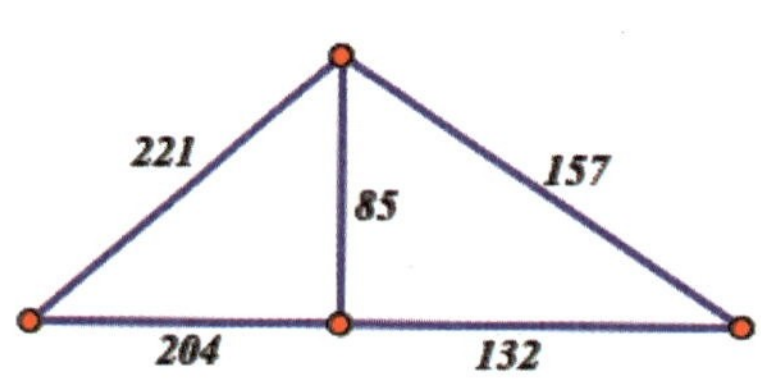

   $P = 204 + 132 + 157 + 221 = 714$
   $A = \frac{1}{2}(204 + 132)(85) = 14280$
   Let's use $A = \frac{1}{2}Pr$ to find the incircle radius.
   $14280 = \frac{1}{2}(714)r$
   $r = 40$

Heronian Triangle 2:

Scale the 5-12-13 triangle by a factor of 11 to get the 55-132-143 triangle. Leave the 85-132-157 triangle as is and put them together as in the figure below:

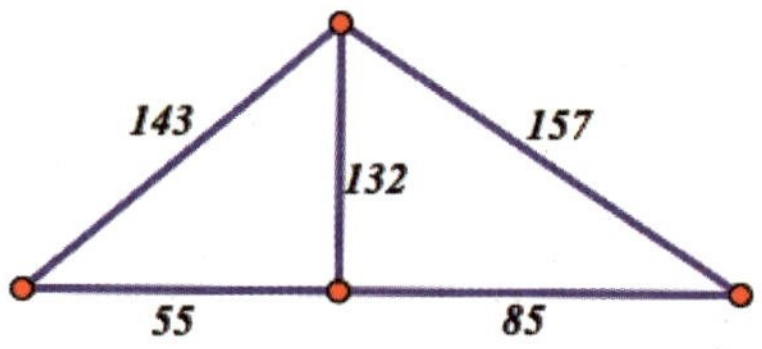

$P = 55 + 85 + 157 + 143 = 440$
$A = \frac{1}{2}(55 + 85)(132) = 9240$
Let's use $A = \frac{1}{2}Pr$ to find the incircle radius.
$9240 = \frac{1}{2}(440)r$
$r = 42$

2. **a.** $x = 3, 17$
Note that $(x - 3)(x - 17) = x^2 - 20x + 51$.

b. $x = \frac{3}{2}, \frac{17}{2}$
Notice that $(2x - 3)(2x - 17) = 4x^2 - 40x + 51$. The coefficient of x^2 from part (a) has been multiplied by 2^2 and the coefficient of x from has been multiplied by 2. Hmmm . . .

c. $x = \frac{3}{10}, \frac{17}{10}$
Notice that $(10x - 3)(10x - 17) = 100x^2 - 200x + 51$. The coefficient of x^2 from part (a) has been multiplied by 10^2 and the coefficient of x from has been multiplied by 10. Hmmm . . .

d. $x = \frac{3}{100}, \frac{17}{100}$
Notice that $(100x - 3)(100x - 17) = 10000x^2 - 2000x + 51$. The coefficient of x^2 from part (a) has been multiplied by 100^2 and the coefficient of x from has been multiplied by 100. Hmmm . . .

3. **a.** $x^2 - 14x + 45 = 0 \Rightarrow (x - 5)(x - 9) = 0 \Rightarrow x = 5, 9$

b. Remember in the previous problem, when we said "Hmmm . . . "? Use those ideas!
$4x^2 - 28x + 45 = (2^2)x - (2)14x + 45 = 0 \Rightarrow x = \frac{5}{2}, \frac{9}{2}$

c. We have $(10^2)x - (10)14x + 45 = 0 \Rightarrow x = \frac{5}{10}, \frac{9}{10}$

d. We have $(100^2)x - (100)14x + 45 = 0 \Rightarrow x = \frac{5}{100}, \frac{9}{100}$

4. **a.** $\left(x - \frac{2}{3}\right)\left(x + \frac{4}{7}\right) = x^2 - \frac{2}{21} - \frac{8}{21}$

b. $x = \frac{2}{3} \Rightarrow 3x - 2 = 0$. $x = -\frac{4}{7} \Rightarrow 7x + 4 = 0$.
$(3x - 2)(7x + 4) = 21x^2 - 2x - 8$ Boom!

c. $\frac{2}{3} + -\frac{4}{7} = \frac{2}{21}$. Notice that 21 is the coefficient of x^2 and 2 is the coefficient of x. Why is this? See the next part . . .

d. If we have a quadratic equation $ax^2+bx+c = 0$, then the solutions are
$x_1 = \frac{-b+\sqrt{b^2-4ac}}{2a}$ and $x_2 = \frac{-b-\sqrt{b^2-4ac}}{2a}$
The sum of the solutions (like we did in part (c)) is $x_1 + x_2 = -\frac{b}{a}$, which contains the coefficients of x and x^2.
The average of the solutions is $\frac{x_1+x_2}{2} = -\frac{b}{2a}$. We recognize this as the x-coordinate of the vertex of the parabola represented by $ax^2 + bx + c$. Since a parabola is symmetric about the vertical line through its vertex, the x-coordinate of the vertex should be halfway in between the x-intercepts (i.e. the average of the solutions to $ax^2 + bx + c = 0$.
Alternatively, you could avoid the quadratic formula completely.
Say that $ax^2 + bx + c = (mx - n)(px - q)$ which has solutions $x = \frac{n}{m}, \frac{q}{p}$ Now

$$\begin{aligned}
&(mx - n)(px - q) \\
&\qquad = (mp)x^2 - (mq + np)x + nq \\
&\qquad \Rightarrow a = mp, b = -(mq + np)
\end{aligned}$$

The sum of the solutions is $\frac{n}{m} + \frac{q}{p} = \frac{np+mq}{mp} = -\frac{b}{a}$
And, the average is $-\frac{b}{2a}$

5. Use $21x^2 - 2x - 8$

   a. The graph is a parabola opening up as illustrated below:

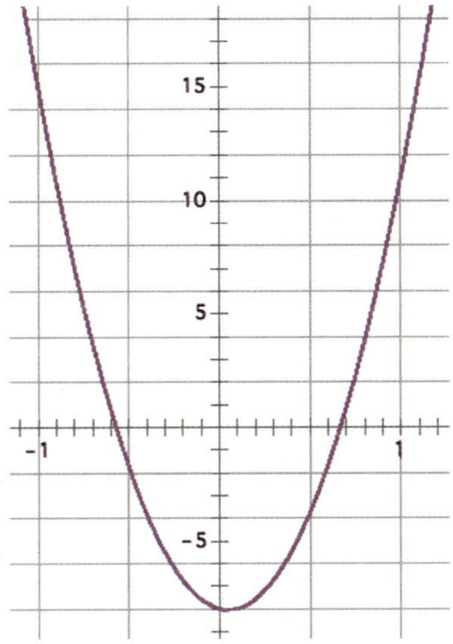

   b. The x-intercepts are $\frac{2}{3}$ and $-\frac{4}{7}$

   c. The x-coordinate of the vertex is $\frac{1}{21}$

6. $x = -\frac{b}{2a}$

7. a. $a(P) = \sqrt{(x-3)^2 + (0-0)^2} = |x-3|$. The graph is below and you can see that the minimum distance is 0 when $x = 3$.

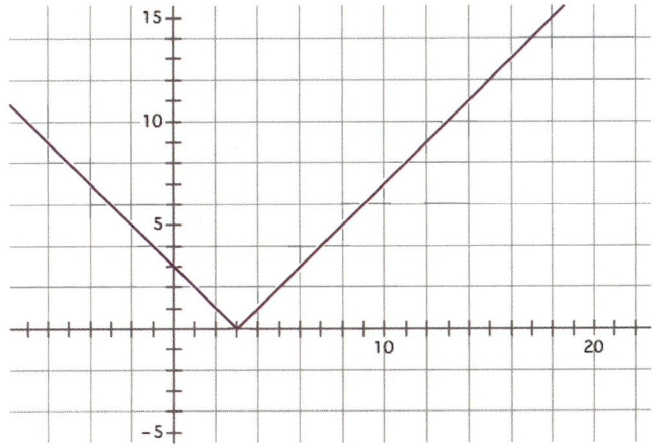

b. $a(P) = |x - 3| + |x - 17|$. The graph is below and you can see that the minimum distance is 14 when $3 \leqslant x \leqslant 17$.

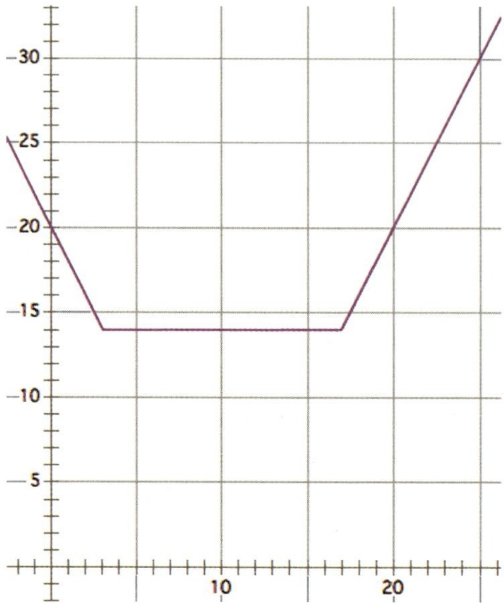

c. $a(P) = |x - 3| + |x - 17| + |x - 1|$. The graph is below and you can see that the minimum distance is 16 when $x = 3$.

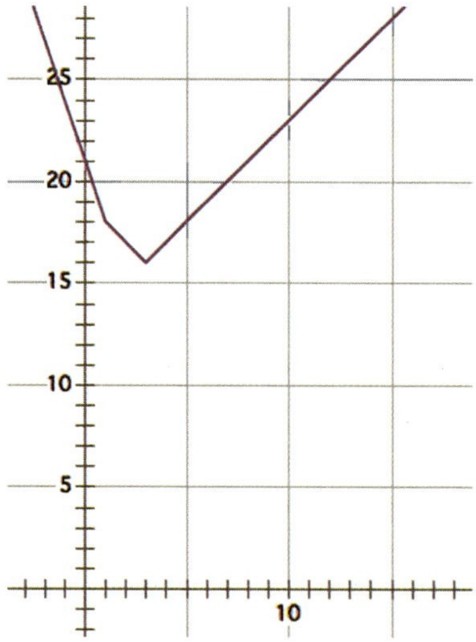

d. $a(P) = |x - 3| + |x - 17| + |x - 1| + |x - 11|$.
The graph is below and you can see that the minimum distance is 24 when $3 \leqslant x \leqslant 11$.

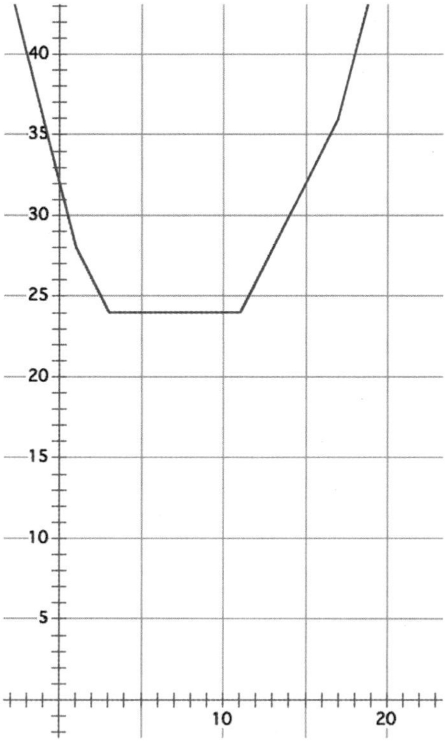

8. **a.** $b(P) = (\sqrt{(x-3)^2 + (0-0)^2})^2 = (x - 3)^2$.
The graph is below and you can see that the minimum distance is 0 when $x = 3$.

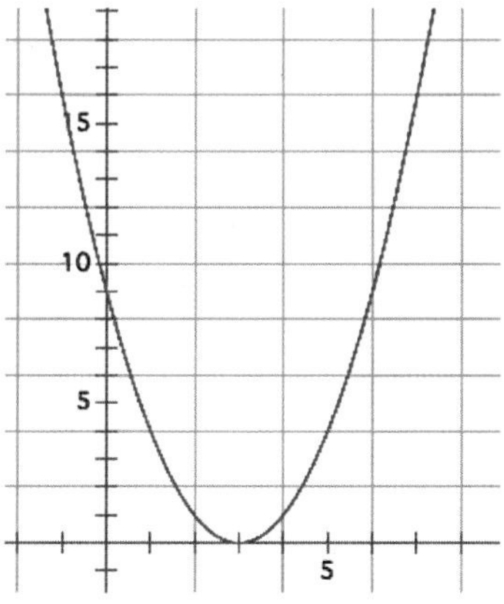

b. $b(P) = (x-3)^2+(x-17)^2$. The graph is below. The minimum distance is 98 when $x = 10$.

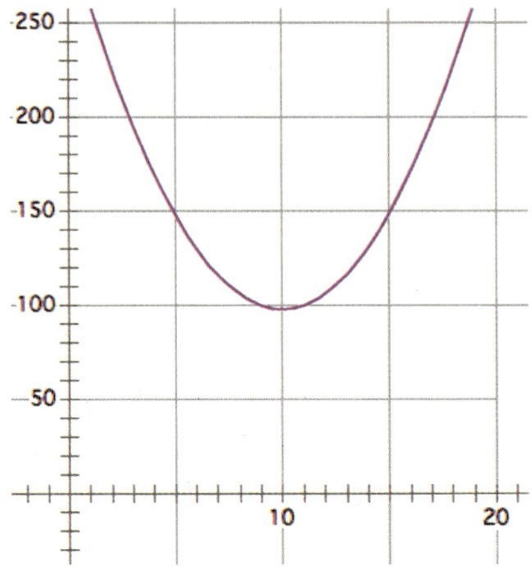

c. $b(P) = (x-3)^2 + (x-17)^2 + (x-1)^2$. The graph is below. The minimum distance is 152 when $x = 7$.

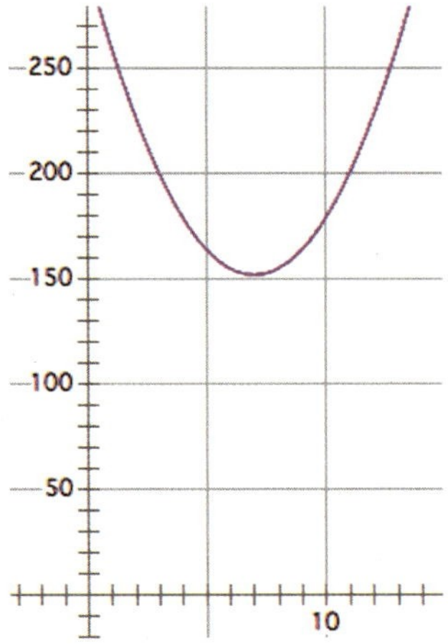

d. $b(P) = (x-3)^2 + (x-17)^2 + (x-1)^2 + (x-11)^2$. The graph is below. The minimum distance is 164 when $x = 8$.

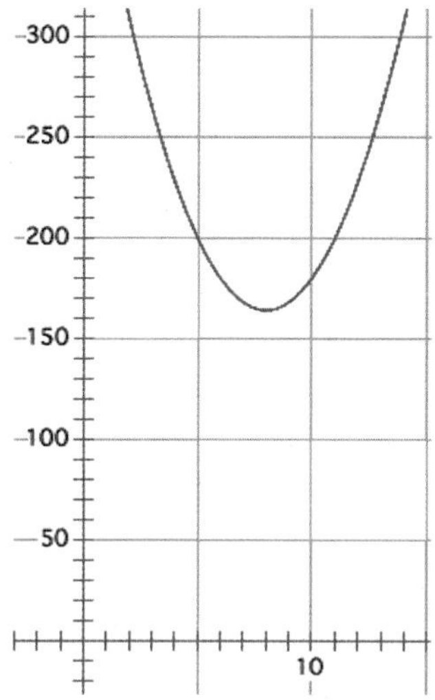

9. **a.** The distance from $(-15, 0)$ to $(-9, 0)$ is 6
The distance from $(-15, 0)$ to $(9, 0)$ is 24
The sum of the distances is indeed 30

b. The distance from $(0, 12)$ to $(-9, 0)$ is 15
The distance from $(0, 12)$ to $(9, 0)$ is 15
The sum of the distances is indeed 30

c. The distance from $(6, 11)$ to $(-9, 0)$ is $\sqrt{346}$
The distance from $(6, 11)$ to $(9, 0)$ is $\sqrt{130}$
The sum of the distances is not 30, though it is close. $\sqrt{346} + \sqrt{130} \approx 30.0028$

d. The distance from $\left(9, -\frac{48}{5}\right)$ to $(-9, 0)$ is $\frac{102}{5}$
The distance from $\left(9, -\frac{48}{5}\right)$ to $(9, 0)$ is $\frac{48}{5}$
The sum of the distances is indeed 30

e. The distance from $(12, b)$ to $(-9, 0)$ is $\sqrt{21^2 + b^2}$
The distance from $(12, b)$ to $(9, 0)$ is $\sqrt{3^2 + b^2}$
In order for the sum of the distances to be 30, we need to satisfy the equation:

$$\sqrt{21^2 + b^2} + \sqrt{3^2 + b^2} = 30$$
$$\sqrt{21^2 + b^2} = 30 - \sqrt{3^2 + b^2}$$
$$21^2 + b^2 = 30^2 - 60\sqrt{3^2 + b^2} + 3^2 + b^2$$
$$60\sqrt{3^2 + b^2} = 30^2 + 3^2 - 21^2$$
$$\sqrt{3^2 + b^2} = \frac{39}{5}$$
$$9 + b^2 = \frac{39^2}{5^2}$$
$$b = \pm\sqrt{\frac{39^2}{5^2} - 9} = \pm\sqrt{\frac{1296}{25}} = \pm\frac{36}{5}$$

In order to be true the points must be $\left(12, \pm\frac{36}{5}\right)$.

f. In order for a point (x, y) to have a combined distance of 30 from both $(-9, 0)$ and $(9, 0)$, the following equation must be satisfied:
$\sqrt{(x+9)^2+y^2}+\sqrt{(x-9)^2+y^2}=30$
$\sqrt{(x+9)^2+y^2}=30-\sqrt{(x-9)^2+y^2}$
$(x+9)^2+y^2=30^2-60\sqrt{(x-9)^2+y^2}$
$+(x-9)^2+y^2$
$x^2+18x+9^2=30^2-60\sqrt{(x-9)^2+y^2}+x^2-$
$18x+9^2$
$60\sqrt{(x-9)^2+y^2}=30^2-36x$
$(x-9)^2+y^2=\left(\frac{30^2-36x}{60}\right)^2$
$(x-9)^2+y^2=\frac{36^2}{60^2}(25-x)^2$
$(x-9)^2+y^2=\frac{3^2}{5^2}(25-x)^2$
$x^2-18x+81+y^2=\frac{9}{25}(x^2-50x+625)$
$\frac{16}{25}x^2+y^2=144$
$\frac{x^2}{225}+\frac{y^2}{144}=1$
All points (x, y) on the ellipse with center $(0, 0)$, major axis vertices $(\pm 15, 0)$, and minor axis vertices $(0, \pm 12)$

10. **a.** True. The distance from $(15, 0)$ to the line $x = 25$ is 10.
The distance from $(15, 0)$ to $(9, 0)$ is 6, which is $\frac{3}{5}$ of 10

b. True. The distance from $(0, -12)$ to the line $x = 25$ is 25
The distance from $(0, -12)$ to $(9, 0)$ is 15, which is $\frac{3}{5}$ of 25.

c. False. The distance from $(13, 6)$ to the line $x = 25$ is 12
The distance from $(13, 6)$ to $(9, 0)$ is $\sqrt{52}$, which is not $\frac{3}{5}$ of 12.

d. True. The distance from $(9, \frac{48}{5})$ to the line $x = 25$ is 16
The distance from $(9, \frac{48}{5})$ to $(9, 0)$ is $\frac{48}{5}$, which is $\frac{3}{5}$ of 16.

e. To be true, the following equation must be satisfied:
$\sqrt{(9+12)^2+b^2}=\frac{3}{5}\sqrt{(25+12)^2+(b-b)^2}$
$\sqrt{21^2+b^2}=\frac{3}{5}\sqrt{37^2}$
$\sqrt{21^2+b^2}=\frac{3}{5}\cdot 37$
$21^2+b^2=\frac{3^2\cdot 37^2}{5^2}$
$b^2=\frac{3^2\cdot 37^2}{5^2}-21^2=\frac{1296}{25}$
$b=\pm\frac{36}{5}$.
The points must be $\left(-12, \pm\frac{36}{5}\right)$ in order to be true.

f. For the distance from a point (x, y) to $(9, 0)$ to be equal to $\frac{3}{5}$ of the distance from (x, y) to the line $x = 25$, the following equation must be satisfied:
$\sqrt{(x-9)^2 + y^2} = \frac{3}{5}\sqrt{(x-25)^2 + (y-y)^2}$
$\sqrt{(x-9)^2 + y^2} = \frac{3}{5}\sqrt{(x-25)^2}$
$(x-9)^2 + y^2 = \frac{3^2}{5^2}(x-25)^2$
$x^2 - 18x + 9^2 + y^2 = \frac{3^2}{5^2}(x^2 - 50x + 25^2)$
$\frac{16}{25}x^2 + y^2 = 144$
$\frac{x^2}{225} + \frac{y^2}{144} = 1$
All points (x, y) on the ellipse with center $(0, 0)$, major axis vertices at $(\pm 15, 0)$, and minor axis vertices at $(0, \pm 12)$ will make the statement true.

11. The ellipse from Problem 10 has center at $(0, 0)$ and focus at $(9, 0)$ so the distance between them is $c = 9$. A vertex is at $(15, 0)$, so the distance to the center is $a = 15$. Thus the eccentricity is $c/a = 9/15 = 3/5$.

12. **a.** True. The distance from $(7, 2)$ to $(9, 0)$ is $\sqrt{(9-7)^2 + 2^2} = 2\sqrt{2}$. The distance from $(7, 2)$ to the line $y = x$ is $|7-2|/\sqrt{2} = \frac{5}{2}\sqrt{2}$. Then $\frac{4}{5}(\frac{5}{2}\sqrt{2}) = 2\sqrt{2}$.

b. True. The distance from $(23, -2)$ to $(9, 0)$ is $\sqrt{(14)^2 + 2^2} = 2\sqrt{50} = 10\sqrt{2}$. The distance from $(23, -2)$ to the line $y = x$ is $|23+2|/\sqrt{2} = \frac{25}{2}\sqrt{2}$. Then $\frac{4}{5}(\frac{25}{2}\sqrt{2}) = 10\sqrt{2}$.

c. True. The distance from $(11, -14)$ to $(9, 0)$ is $\sqrt{2^2 + 14^2} = 2\sqrt{50} = 10\sqrt{2}$. The distance from $(11, -14)$ to the line $y = x$ is $|11 + 14|/\sqrt{2} = \frac{25}{2}\sqrt{2}$. Then $\frac{4}{5}(\frac{25}{2}\sqrt{2}) = 10\sqrt{2}$.

d. False. The distance from $(19, -19)$ to $(9, 0)$ is $\sqrt{10^2 + 19^2} = \sqrt{461}$. The distance from $(19, -19)$ to the line $y = x$ is $19\sqrt{2}$. Then $\frac{4}{5}19\sqrt{2} \neq \sqrt{461}$. They cannot be exactly equal since 2 does not divide the prime number 461. However, the $\sqrt{461}$ is very close to the other quantity, about .025 less.

e. True. The distance from $(27, -18)$ to $(9, 0)$ is $\sqrt{18^2 + 18^2} = 18\sqrt{2}$. The distance from $(27, -18)$ to the line $y = x$ is $|27 + 18|/\sqrt{2} = \frac{45}{2}\sqrt{2}$. Then $\frac{4}{5}(\frac{45}{2}\sqrt{2}) = 18\sqrt{2}$.

f. The distance from the point (x, y) to the point $(9, 0)$ is $\sqrt{(x-9)^2+y^2}$. The distance from the point (x, y) to the line with equation $y = x$ is $\frac{|x-y|}{\sqrt{2}}$. The requirement on (x, y) is that

$$\sqrt{(x-9)^2+y^2} = \frac{4}{5}\frac{|x-y|}{\sqrt{2}}$$

Squaring both sides to get a polynomial equation, this becomes

$$(x-9)^2+y^2 = \frac{8}{25}(x-y)^2$$

13. The expansion of one expression is

$$(a^2+b^2)(c^2+d^2) = a^2c^2+a^2d^2+b^2c^2+b^2d^2$$

and the expansion of th other is

$$\begin{aligned}&(ac-bd)^2+(ad+bc)^2\\ &\quad = a^2c^2-2acbd+b^2d^2+a^2d^2+2adbc+b^2c^2\\ &\quad = a^2c^2+b^2d^2+a^2d^2+b^2c^2\end{aligned}$$

Since these are equal, the original expressions are equivalent expressions.

14. It follows directly from Problem 13 that if $m = a^2+b^2$ and $n = c^2+d^2$, then $mn = (a^2+b^2)(c^2+d^2) = (ac-bd)^2+(ad+bc)^2$. Thus, setting $p = ac-bd$ and $q = ad+bc$, then $mn = p^2+q^2$.

15. Let $z = a+ib$ and $w = c+id$. Then $|z|^2|w|^2 = (a^2+b^2)(c^2+d^2)$. Since $zw = (ac-bd)+(ad+bc)i$, then $|zw|^2 = (ac-bd)^2+(ad+bc)^2$. From Problem 13 we see that the magnitude squared $|z|^2|w|^2 = |zw|^2$. Taking the square root of these nonnegative numbers implies that $|z||w| = |zw|$.

16. Let $z = a+ib$ and $w = c+id$. If $z+w = (a+c)+(b+d)i$ is real, then $b+d = 0$. This says $z = a+ib$ and $w = c-ib$. The product $zw = (ac+b^2)+(-ab+bc)i$. If this is real, then $b(c-a) = 0$. This is true only if either $b = 0$ or $c = a$. If $b = 0$, then $z = a+i0 = a$ and $w = c+0i = c$ are real. If $c = a$, then $z = a+ib$ and $w = a-ib = \bar{z}$.

17. To find the eccentricity of the shape in Problem 12, we first find the vertices. By symmetry, these must lie on the line through $(9,0)$ that is perpendicular to the line with equation $y = x$. The equation for this line is $y = 9 - x$. If we substitute a point on this line $x, 9 - x$ into the quadratic equation in Problem 12 and solve for x, we get two points: $(x, 9-x) = (7,2)$ and $(x, 9-x) = (27,-18)$, which are points (a) and (e) of Problem 12.

The center of the shape is $(\frac{7+27}{2}, \frac{2-18}{2}) = (17,-8)$, the midpoint of these two points.

The distance from the center to the focus $(9,0)$ is $c = \sqrt{(17-9)^2 + (-8-0)^2} = \sqrt{8^2+8^2} = 8\sqrt{2}$.

The distance from the center to the vertex $(7,2)$ is $a = \sqrt{(17-7)^2 + (-8-2)^2} = \sqrt{10^2+10^2} = 10\sqrt{2}$.

Thus the eccentricity $e = \frac{c}{a} = \frac{8\sqrt{2}}{10\sqrt{2}} = \frac{4}{5}$.

Problem Set 11

Opener

Well, here is the data. Where do **you** want the line?

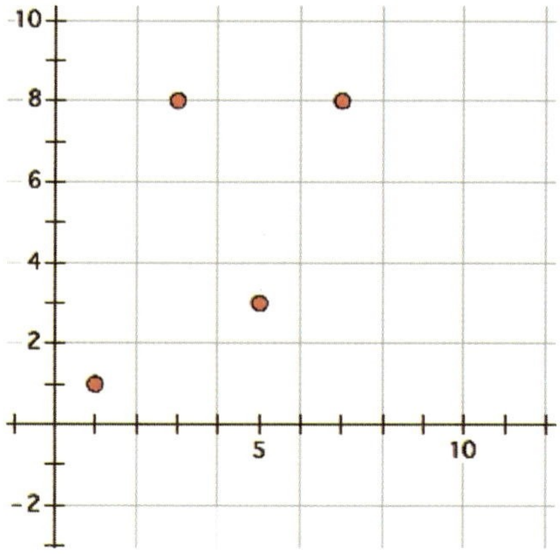

1. **a.** Trevor used $y = x$, which gives the following ordered paris: $(1,1), (3,3), (5,5), (7,7)$. The sum of squared errors is:
$(1-1)^2 + (8-3)^2 + (3-5)^2 + (8-7)^2 = 0 + 25 + 4 + 1 = 30$

b. Marakina used $y = x - 1$, which gives the following ordered paris: $(1,0),(3,2),(5,4),(7,6)$.
The sum of squared errors is:
$(1-0)^2 + (8-2)^2 + (3-4)^2 + (8-6)^2 = 1 + 36 + 1 + 4 = 42$
Olimpia used $y = x + 1$, which gives the following ordered paris: $(1,2),(3,4),(5,6),(7,8)$.
The sum of squared errors is:
$(1-2)^2 + (8-4)^2 + (3-6)^2 + (8-8)^2 = 1 + 16 + 9 + 0 = 26$

c. Using $y = x + b$, we get the ordered pairs: $(1,1+b),(3,3+b),(5,5+b),(7,7+b)$.
The sum of squared errors is then:
$(1-(1+b))^2 + (8-(3+b))^2 + (3-(5+b))^2 + (8-(7+b))^2$
$= (-b)^2 + (5-b)^2 + (-2-b)^2 + (1-b)^2$
$= 4b^2 - 8b + 30$
This is a quadratic, which has a minimum at its vertex. The vertex of the parabola is $-\frac{b}{2a} = \frac{8}{2\cdot 4} = 1$. $b = 1$ gives $y = x + 1$ as the best possible line of the form $y = x + b$.

d. See previous part.

2. Use $y = mx + b$ for each part, and we want to find the best line for each slope

a. $y = 2x + b$ gives the ordered pairs: $(1,2+b),(3,6+b),(5,10+b),(7,14+b)$. The sum of squared errors is:
$(1-(2+b))^2 + (8-(6+b))^2 + (3-(10+b))^2 + (8-(14+b))^2 = 4b^2 + 24b + 90$, which is a parabola with a minimum at its vertex. So $b = -\frac{24}{8} = -3$ gives the best line for a slope of 2.
Equation of the line: $y = 2x - 3$

b. $y = 3x + b$ gives the ordered pairs: $(1,3+b),(3,9+b),(5,15+b),(7,21+b)$. The sum of squared errors is:
$(1-(3+b))^2 + (8-(9+b))^2 + (3-(15+b))^2 + (8-(21+b))^2 = 4b^2 + 56b + 318$, which is a parabola with a minimum at its vertex. So $b = -\frac{56}{8} = -7$ gives the best line for a slope of 3.
Equation of the line: $y = 3x - 7$

c. $y = 4x + b$ gives the ordered pairs: $(1,4+b),(3,12+b),(5,20+b),(7,28+b)$. The sum of squared errors is:
$(1-(4+b))^2 + (8-(12+b))^2 + (3-(20+b))^2 + (8-(28+b))^2 = 4b^2 + 88b + 714$, which is

a parabola with a minimum at its vertex. So $b = -\frac{88}{8} = -11$ gives the best line for a slope of 4.

Equation of the line: $y = 4x - 11$

d. $y = 0x + b$ gives the ordered pairs: $(1, 0 + b), (3, 0 + b), (5, 0 + b), (7, 0 + b)$. The sum of squared errors is:
$(1 - b)^2 + (8 - b)^2 + (3 - b)^2 + (8 - b)^2 = 4b^2 - 40b + 138$, which is a parabola with a minimum at its vertex. So $b = \frac{40}{8} = 5$ gives the best line for a slope of 0.

Equation of the line: $y = 5$

e. $y = -1x + b$ gives the ordered pairs: $(1, -1 + b), (3, -3 + b), (5, -5 + b), (7, -7 + b)$. The sum of squared errors is:

$$\begin{aligned}(1 - (-1 + b))^2 &+ (8 - (-3 + b))^2 \\ &+ (3 - (-5 + b))^2 + (8 - (-7 + b))^2 \\ = 4b^2 - 72b + 414,\end{aligned}$$

which is a parabola with a minimum at its vertex. So $b = \frac{72}{8} = 9$ gives the best line for a slope of -1.

Equation of the line: $y = -x + 9$

3. See the graphs below:

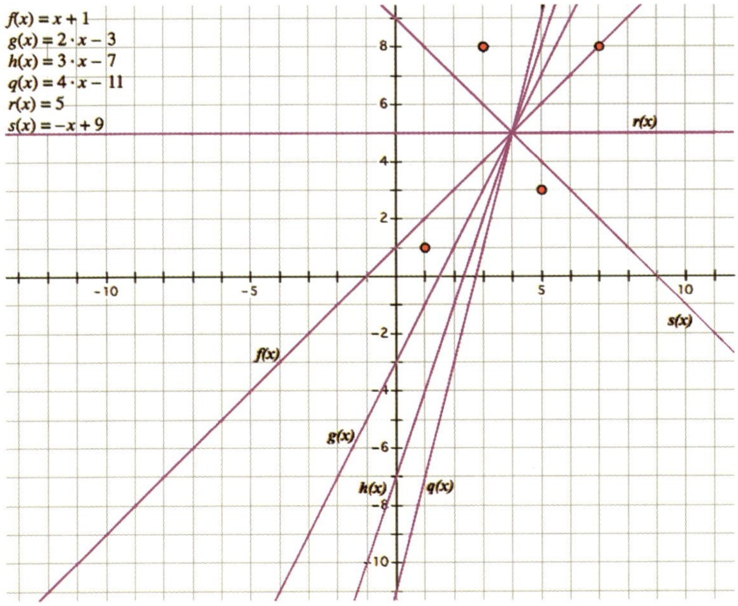

Notice that all lines are passing through the point $(4,5)$. The mean of the x-coordinates of the data is 4. The mean of the y-coordinates of the data is 5. Whoa!
So the best line will be of the form $y-5=m(x-4)$ for each m.

4. Now we are using $y = mx$ with new data. Our new ordered pairs with this line are: $(-3,-3m)$, $(-1,-m)$, $(1,m)$, and $(3,3m)$. The sum of squared errors is $(-4+3m)^2+(3+m)^2+(-2-m)^2+(3-3m)^2 = 20m^2-32m+38$, which is a parabola with a minimum at its vertex. So $m = \frac{32}{40} = \frac{4}{5}$ gives the best slope for a line of the form $y = mx$. Best m ever!
Equation of the line: $y = \frac{4}{5}x$

5.

a. Hopefully your trusty calculator gives a linear regression line of $y = \frac{4}{5}x$. The slope is $\frac{4}{5}$

b. The graph passes of the linear regression line passes through the point $(0,0)$. The x-coordinates of the data have a mean of 0, as do the y-coordinates of the data. So, as before, the best line will be of the form $y = mx$.

c. Now to find the sum of squared errors (a.k.a. *the badness*). The line $y = \frac{4}{5}x$ gives the ordered pairs:
$\left(-3,-\frac{12}{5}\right), \left(-1,-\frac{4}{5}\right), \left(1,\frac{4}{5}\right), \left(3,\frac{12}{5}\right)$
The sum of squared errors is:
$\left(-4+\frac{12}{5}\right)^2+\left(3+\frac{4}{5}\right)^2+\left(-2-\frac{4}{5}\right)^2+\left(3-\frac{12}{5}\right)^2 = \frac{126}{5}$
We have already shown that this is the *best* line.

6. Make sure you are using the data from the box!

a. Lauren's ordered pairs from using $y = x$ are: $(1,1), (3,3), (5,5), (7,7)$. The sum of absolute errors is: $|1-1|+|8-3|+|3-5|+|8-7| = 8$

b. James's ordered pairs from using $y = x-1$ are: $(1,0), (3,2), (5,4), (7,6)$. The sum of absolute errors is: $|1-0|+|8-2|+|3-4|+|8-6| = 10$
Timon's ordered pairs from using $y = x+1$ are: $(1,2), (3,4), (5,6), (7,8)$. The sum of absolute errors is: $|1-2|+|8-4|+|3-6|+|8-8| = 8$
Timon did better than James.

c. Allison's ordered pairs from using $y = x+\frac{1}{2}$ are: $\left(1,\frac{3}{2}\right), \left(3,\frac{7}{2}\right), \left(5,\frac{11}{2}\right), \left(7,\frac{15}{2}\right)$. The sum of absolute errors is: $\left|1-\frac{3}{2}\right|+\left|8-\frac{7}{2}\right|+\left|3-\frac{11}{2}\right|+\left|8-\frac{15}{2}\right| = 8$
Allison did as well as Timon and Lauren.

d. Using $y = x + b$ we get the ordered pairs:
$(1, 1+b), (3, 3+b), (5, 5+b), (7, 7+b)$.
The sum of absolute errors is:
$|-b| + |5-b| + |-2-b| + |1-b|$
$= |b| + |5-b| + |2+b| + |1-b|$

e. The graph of $f(b) = |b|+|5-b|+|2+b|+|1-b|$ is displayed below:

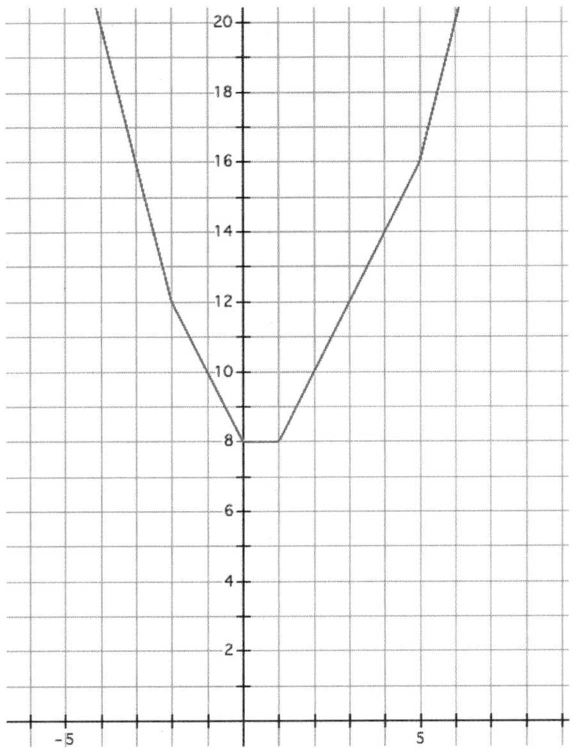

Notice that the graph is similar in shape to a parabola. Also notice that for $0 \leqslant b \leqslant 1$ will result in the equation $y = x + b$ giving a minimum value for the sum of absolute errors, which is why Lauren, Timon, and Allison had equally good lines.

The mathematical ideas here are the same ones used for *One-Dimensional Darryl* in Problem Set 3 when we were working his round-trip problem.

7. To find the other solution by means of the sketch, we drag the radius of the circle so that the image curve passes through the point $6 + i0$. Then we drag the point z on the circle until the image point w is at $6 + i0$. This happens when $z = -3$. We can check directly that this is the other solution of the equation.

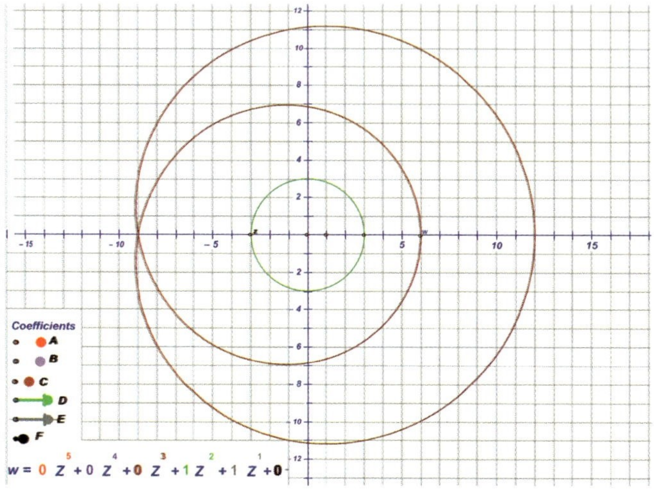

8. To solve these equation, we drag the slider so that the image curve is the image of the circle by the map $z^2 - z$. Then we solve for the various values as in Problem 7.

   a. Setting $z = -2$ moves w to 6. The other solution (not shown) is $z = 3$.

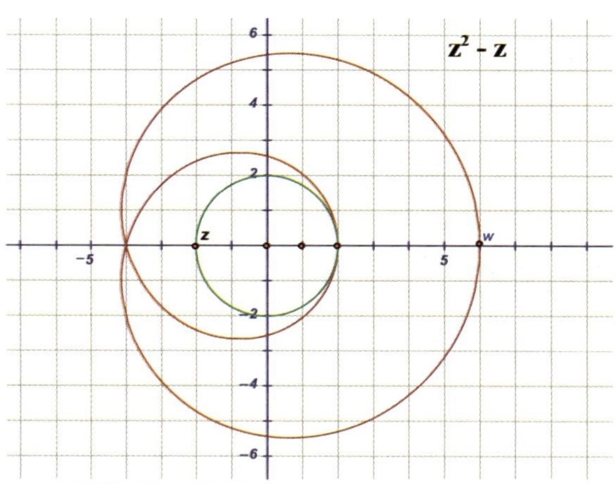

b. Moving the point z to $z = 4$ moves w to 12. The other solution (not shown) is $z = -3$.

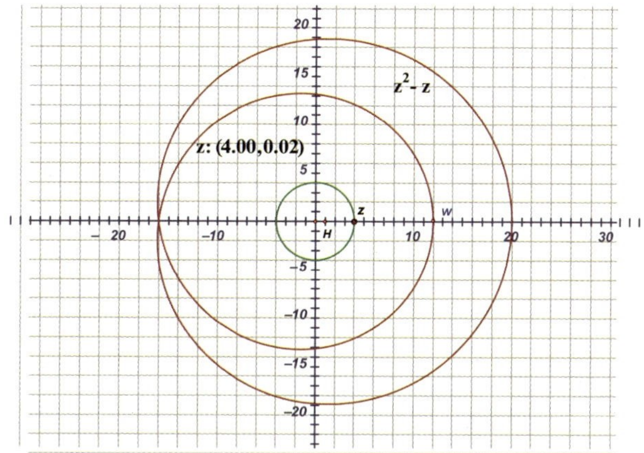

c. Moving z to the point $\frac{1}{2} + 3i$ gives an approximate solutions. The conjugate point $z = \frac{1}{2} - 3i$ is also an approximate solution.

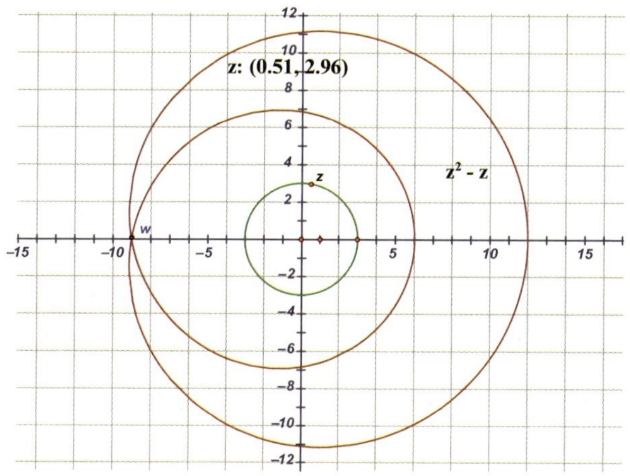

d. Moving the point z to $z = -4$ moves w to 20. The other solution (not shown) is $z = 5$.

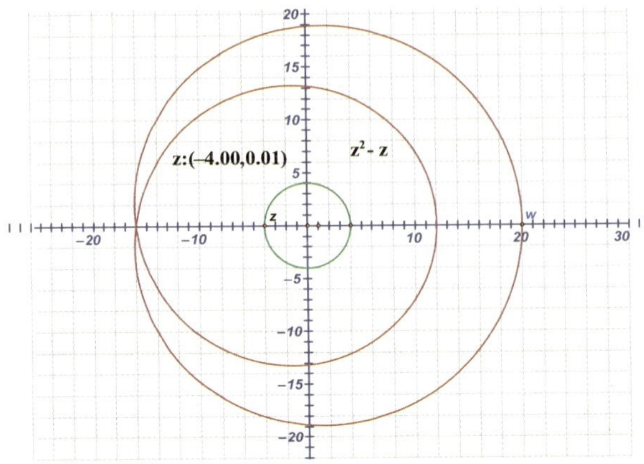

9. The conjugate of $z = x + yi$ is $\overline{z} = x - yi$, which is the reflection of z in the x-axis.

The conjugate of $z^2 + z$ is $\overline{z^2 + z} = \overline{z}^2 + \overline{z}$ since $\overline{zw} = \overline{z}\,\overline{w}$ for any $z = x + yi$ and $w = u + vi$. To see this, multiply and expand the produces on each side of the equation.

Notice that in the Sketchpad figure the image of the circle is symmetric about the x-axis.

10. This equation says that the distance from (x, y) to the point $(25, 0)$ is 30 more than the distance from (x, y) to the point $(-25, 0)$.

$$\sqrt{(x-25)^2 + y^2} = 30 + \sqrt{(x+25)^2 + y^2}$$

This can be converted to a quadratic polynomial equation by squaring two times. The first time gives

$$(x-25)^2+y^2 = 900+60\sqrt{(x+25)^2 + y^2}+(x+25)^2+y^2$$

Expanding and rearranging so that the square root is alone on one side gives

$$\begin{aligned}(x-25)^2 - (x+25)^2 - 900 &= -100x - 900\\ &= 60\sqrt{(x+25)^2 + y^2}\end{aligned}$$

Dividing both sides by 10 and then squaring gives

$$\begin{aligned}&100x^2 + 1800x + 8100\\ &\qquad = (10x+90)^2 = 36((x+25)^2 + y^2)\\ &\qquad = 36x^2 + 1800x + 22500 + 36y^2\end{aligned}$$

Collecting and rearranging finally gives

$$64x^2 - 36y^2 = 120^2 \quad \text{or} \quad (\frac{x}{3})^2 - (\frac{y}{4})^2 = 5^2$$

Since we squared the original equation, while we can say that every point (x, y) satisfying the original equation satisfies the new one, we cannot say that every solution of the polynomial equation satisfies the original. For example, the point $(15, 0)$ satisfies the second equation but not the first.

If we replace the original equation with

$$\sqrt{(x + 25)^2 + y^2} = 30 + \sqrt{(x - 25)^2 + y^2}$$

and proceed as above, we get the same quadratic polynomial equation. The point $(15, 0)$ does satisfy this equation. It seems that the solution set of the quadratic polynomial equation consists of two disjoint parts, each described by one of the distance equations.

The two distance equations can be combined into one:

$$|\sqrt{(x + 25)^2 + y^2} - \sqrt{(x - 25)^2 + y^2}| = 30$$

11. This set is the set of points (x, y) such that

$$\sqrt{(x - 25)^2 + y^2} = \frac{5}{3}|x - 9|$$

Squaring both sides,

$$(x - 25)^2 + y^2 = \frac{25}{9}(x - 9)^2$$

Expanding gives

$$x^2 - 50x + 625 + y^2 = \frac{25}{9}x^2 - 50x + 225$$

Collecting,

$$\frac{16}{9}x^2 - y^2 = 400 \quad \text{or} \quad (\frac{x}{3})^2 - (\frac{y}{4})^2 = 5^2$$

This is the same equation as in Problem 10. Unlike the previous problem, the original equation of distances is satisfied both by $(15, 0)$ and $(-15, 0)$. Thus the original equation includes both parts of the hyperbola.

12. The hyperbolas in Problems 10 and 11 are the same. The center of this hyperbola is $(0, 0)$, one focus is $(25, 0)$ and a vertex is $(15, 0)$. Thus the number $c = 25$ and the number $a = 15$, so the eccentricity $e = 25/15 = 5/3$.

13. For David's data, the total absolute error for any line $y = mx + b$ can be computed. Label the errors e_1, e_2, e_3, e_4. Then

$$\begin{aligned} e_1 &= |1-(m+b)| = |(1-m)-b| \\ e_2 &= |8-(3m+b)| = |(8-3m)-b| \\ e_3 &= |3-(5m+b)| = |(3-5m)-b| \\ e_4 &= |8-(7m+b)| = |(8-7m)-b| \end{aligned}$$

Then for any fixed m the graph of he error as a function of b is the sum of four absolute value functions $|y_i - mx_i - b|$, each with slope -1 for $b < |y_i - mx_i|$ and slope + 1 for $b > |y_i - mx_i|$.

A function which is the sum of four such absolute value functions has slope -4 for very negative values of b and slope +4 for very positive values. In between for intervals the slope is -2, 0, +2, with the possibility that the the length of an interval can be zero. The points on the interval with slope 0 or the point between the slopes of -2 and + 2 are/is the minimum error for this value of m.

Thus the minimum values of b occur when b is greater than two of the numbers $(1-m)$, $(8-3m)$, $(3-5m)$, $(8-7m)$ and less than two of them. This means that the sum of the absolute values of the e_i is actually the sum of the greater two of $(1-m)$, $(8-3m)$, $(3-5m)$, $(8-7m)$ less the sum of the smaller two. This is true since whenever $p < b < q$, the sum $|b-p| + |b-q|$ is just the difference $q - p$.

How can we find the minimum values of m and b? One experimental approach is to construct a dynamic graph with Sketchpad. If one fixes the value of m and examines the graph as a function of b, one can see the minimum interval. Then if one varies the graph by dragging the value of$m$ on observes that the lowest minimum level appears to occur when the interval of zero slope is reduced to a point. This seems to happen at about $m = 1.16$. For this value of m the minimum value of b seems to occur at $b = -0.16$.

Is there some reasoning behind this? The ordering of the values of $(1-m)$, $(8-3m)$, $(3-5m)$, $(8-7m)$ will change when two of the values are equal. Looking into this, it suggests that the choice of m with smallest minimum value will occur then. In this case, if we set $(1-m) = (8-7m)$, we get $m = 7/6$, which is about equal to the experimental value. For this m, $b = 1 = m = -1/6$. This also agrees with experiment. For these values of m and b, the minimum total error is $22/3$.

14. For Brian's data, the total error for any line $y = mx+b$ can be computed. Label the errors e_1, e_2, e_3, e_4, e_5. Then

$$\begin{aligned}
e_1 &= |3 - m - b)| \\
e_2 &= |7 - m - b| \\
e_3 &= |5 - 3m - b| \\
e_4 &= |3 - 5m - b| \\
e_5 &= |7 - 5m - b|
\end{aligned}$$

Again, creating a graph for the total absolute error as a function of b so that the graph can be changed dynamically by varying m, we can find the values of m for which the minimum value of the graph is itself minimized.

This time it appears that for all values of $-1 \leqslant m \leqslant 1$ the minimum of the graph is the same, with the value of 8 at the minimum. The values of minimum b appear to be $b = 8$ for $m = -1$, $b = 5$ for $m = 0$, and $b = 2$ for $m = 1$. The value of b appears to vary as a linear function of m.. This leads to a conjecture that the minimum error will be 8 overall and that this is achieved for $-1 \leqslant m \leqslant 1$ and $b = 5 - 3m$.

The errors, which are absolute values, can be expressed as signed numbers for $-1 \leqslant m \leqslant 1$ and $b = 5 - 3m$.

$$\begin{aligned}
e_1 &= |3 - 1m - b| = |3 - m - (5 - 3m)| \\
&= |-2 + 2m| = 2 - 2m \\
e_2 &= |7 - 1m - b| = |7 - m - (5 - 3m)| \\
&= |2 + 2m| = 2 + 2m \\
e_3 &= |5 - 3m - b| = |5 - 3m - (5 - 3m)| = 0 \\
e_4 &= |3 - 5m - b| = |3 - 5m - (5 - 3m)| \\
&= |-2 - 2m| = 2 + 2m \\
e_5 &= |7 - 5m - b| = |7 - 5m - (5 - 3m)| \\
&= |2 - 2m| = 2 - 2m
\end{aligned}$$

Adding these five terms, the total error is 8. Thus all the $-1 \leqslant m \leqslant 1$and $b = 5 - 3m$ define lines with the same total absolute error for this data.

15. Let $z = 34 + i$ and $w = 55 + i$. then $zw = (34 \times 55 - 1) + (34 + 55)i = 1869 + 89i = 89(21 + i)$. This seems surprising.

16. The product $abcd$ is

$$\begin{aligned}
(2 + i)&(5 + i)(13 + i)(21 + i) \\
&= (9 + 7i)(13 + i)(21 + i) \\
&= (110 + 100i)(21 + i) \\
&= 2210 + 2210i = 2210(1 + i)
\end{aligned}$$

The direction is $\pi/4$ radians or 45 degrees.

17. Since the complex number i has direction 90 degrees, then – since cubing a complex number triples the direction – one cube root of i should be the complex number w on the unit circle with direction 30 degrees. Since the right triangle with angle 30 degrees is one-half of an equilateral triangle, one can compute that $w = (\sqrt{3} + i)/2$. Computing the cube, one sees that $w^3 = i$.

If this number is multiplied by a cube root of 1, then the cube of this product should be $1 \times i = i$. What are the cube roots of 1? Using the same reasoning as before, since the direction of 1 can be viewed either as 0 or 360 degrees, the directions of the cube roots should be 0, 120 and 240 degrees, since tripling any of these angles is an integer multiple of 360.

The cube root with 0 degree direction is just 1 itself.

The cube root u with 120 degree direction makes a 30 degree angle with the positive y-axis, so its coordinates are $u = (-1 + i\sqrt{3})/2$. The cube root with 240 degree direction is the conjugate $\overline{u} = (-1 - i\sqrt{3})/2$.

These three complex cube roots of 1 form an equilateral triangle inscribed in the unit circle.

The product

$$\begin{aligned} uw &= (-\frac{1}{2} + i\frac{\sqrt{3}}{2})(\frac{\sqrt{3}}{2} + i\frac{1}{2}) \\ &= (-\frac{\sqrt{3}}{4} - \frac{\sqrt{3}}{4}) + i(\frac{3}{4} - \frac{1}{4}) \\ &= -\frac{\sqrt{3}}{2} + i\frac{1}{2} \end{aligned}$$

$$\begin{aligned} \overline{u}w &= (-\frac{1}{2} - i\frac{\sqrt{3}}{2})(\frac{\sqrt{3}}{2} + i\frac{1}{2}) \\ &= (-\frac{\sqrt{3}}{4} + \frac{\sqrt{3}}{4}) + i(-\frac{3}{4} - \frac{1}{4}) \\ &= 0 - 1i = -i \end{aligned}$$

These three complex cube roots of i form an equilateral triangle inscribed in the unit circle.

For trigonometry fans, since the complex number i has direction $\pi/2$,the direction of w is $\pi/6$ and $w = (\cos\pi/6, i\sin\pi/6)$.

The cube roots of 1 are $1 = (\cos 0\pi, i\sin 0\pi)$, $u = (\cos 2\pi/3, \sin 2\pi/3)$ and $\overline{u} = (\cos 4\pi/3, \sin 4\pi/3)$

Forming these products, we have $w = (\cos\pi/6, i\sin\pi/6)$, $uw = (\cos 5\pi/6, i\sin 5\pi/6)$, and $\overline{u}w = (\cos 9\pi/6, i\sin 9\pi/6) = -i$. These are all among the vertices of a regular dodecagon inscribed in the unit circle.

18. **a.**

$$\begin{aligned} (a+bi)^3 &= a^3 + 3a^2bi - 3ab^2 - b^3i \\ &= (a^3 - 3ab^2) + i(3a^2b - b^3) \end{aligned}$$

b. The complex number $\cos\theta + i\sin\theta$ has direction θ and magnitude 1, since $\cos^2\theta + \sin^2\theta = 1$.

c. The direction of $(\cos\theta + i\sin\theta)^3$ is 3θ and the magnitude is $1^3 = 1$ so

$$\begin{aligned} \cos 3\theta + i\sin 3\theta &= (\cos\theta + i\sin\theta)^3 \\ &= (\cos^3\theta - 3\cos\theta\sin^2\theta) \\ &\quad + i(3\cos^2\theta\sin\theta - \sin^3\theta) \end{aligned}$$

d. Since $\cos^2\theta + \sin^2\theta = 1$, $\cos 3\theta = \cos^3\theta - 3\cos\theta(1-\cos^2\theta) = 4\cos^3\theta - 3\cos\theta$.
Also, $\sin 3\theta = 3(1-\sin^2\theta)\sin\theta - \sin^3\theta = 3\sin\theta - 4\sin^3\theta$.

19. **a.**

$$\begin{aligned}(x+2)^3 &= x^3 + 3(2x^2) + 3(4x) + 8\\ &= x^3 + 6x^2 + 12x + 8\end{aligned}$$

b. A cube has one solid, 6 square faces, 12 edges and 8 vertices. These number match up with the numbers in (a).

c. In lower dimensions: A square has one square, 4 edges and 4 vertices, matching $(x+2)^2 - x^2 + 4x + 4$. An interval has one segment and two endpoints, matching $(x+2)^1 = x+2$. A point has one vertex, matching $(x+2)^0 = 1$.
In higher dimensions

$$\begin{aligned}(x+2)^4 &= x^4 + 4(2x^3) + 6(4x^2) + 4(8x) + 16\\ &= x^4 + 8x^3 + 24x^2 + 32x + 16\end{aligned}$$

This does match the hypercube of dimension 4. In general one way of thinking about extending an n-dimensional cube to an $(n+1)$-dimensional cube is to extrude a segment or take the product with a segment. This means that there are two "ends", each of which is an n-dimensional cube. The "middle" between the two ends consists of all the elements of the original n-dimensional cube stretched to be one dimension higher, so points become segments, segments become squares, squares, become cubes, etc.
The same process happens when multiplying by $(x+2)$. The product with 2 produces two each of the original objects, and the product with x raises the dimension of each original object – just like the geometric description.

Problem Set 12

Opener

So we have this data:

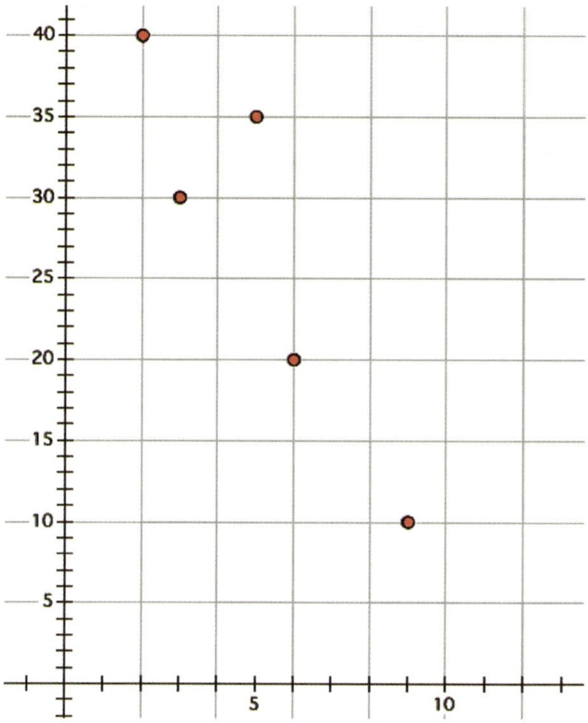

The line of best fit will go through the "balance point" of data, found by calculating the average of the d and n values. The balance point is $(5, 27)$. The equation of the line of best fit is $n - 27 = m(d - 5)$. Find the best value of m by calculating the error as a function of m and minimizing. The best fit occurs when $m = -4$, so the line of best fit is $n - 27 = -4(d - 5)$, or $n = -4d + 47$. The "badness" of this line, the sum of squared errors, is 100.

1. **a.** The sum of squared errors when using the line $y = mx + b$ for the given data will be: $(40 - (2m + b))^2 + (30 - (3m + b))^2 + (35 - (5m+b))^2+(20-(6m+b))^2+(10-(9m+b))^2$

   b. Notice that the mean of the d-values is 5 and the mean of the n-values is 27, both of these values are contained in the expression. Notice also that it is quadratic in both m and b, and the second quantity is such that we can choose values of m that will make solving for b not too bad.

2. $5(b+5m-27)^2+30(m+4)^2+100=130$
$5(b+5m-27)^2+30(m+4)^2=30$
$\frac{(b+5m-27)^2}{6}+(m+4)^2=1$
The easy choice would be to choose $m=-4$ and solve for b:
$\frac{(b+5(-4)-27)^2}{6}+(-4+4)^2=1$
$\frac{(b-20-27)^2}{6}=1$
$(b-47)^2=6$
$b^2-94b+2203=0 \Rightarrow b=47\pm\sqrt{6}$ But these values seem messy. Let's find some others.
Notice if we choose $m=-5$ we end up with a quadratic equal to 0 right away.
$\frac{(b+5(-5)-27)^2}{6}+(-5+4)^2=1$
$\frac{(b-52)^2}{6}+1=1$
$(b-52)^2=0 \Rightarrow b=52$
Now choose $m=-3$, which, again, reduces the equation to a quadratic equal to 0 right away.
$\frac{(b+5(-3)-27)^2}{6}+(-3+4)^2=1$
$\frac{(b-42)^2}{6}+1=1$
$(b-42)^2=0 \Rightarrow b=42$
So our "nice" lines with *badness* of 130 are
$y=-5x+52$ and $y=-3x+42$

3. First of all, the largest A can be is if $A=B=C$, and the only possibility for that is if $\frac{1}{A}+\frac{1}{B}+\frac{1}{C}=\frac{1}{A}+\frac{1}{A}+\frac{1}{A}=\frac{3}{A}=\frac{1}{2}\Rightarrow 6=A=B=C$.
Now let's see what other possibilities are in store for us . . .
$A\leqslant B\leqslant C\Rightarrow \frac{1}{A}\geqslant\frac{1}{B}\geqslant\frac{1}{C}$
To begin with, we can see that $A\geqslant 3$, since $\frac{1}{A}+\frac{1}{B}+\frac{1}{C}=\frac{1}{2}$.
Let's start with $A=3$:
Then we have $\frac{1}{B}+\frac{1}{C}=\frac{1}{6}\Rightarrow\frac{6B}{B-6}=C$.
C must be an integer and we can see that, for this case, we need $B>6$. Running through values of B we get the following: $B=7, C=42$, $B=8, C=24$, $B=9, C=18$, $B=10, C=15$, $B=12, C=12$. Why did we stop at $B=12$? Try a value larger than 12. What happens to C? Is $B\leqslant C$ still satisfied? Nope!

Now let's look at $A = 4$:
Then we have $\frac{1}{B} + \frac{1}{C} = \frac{1}{4} \Rightarrow \frac{4B}{B-4} = C$.
C must be an integer and we can see that, for this case, we need $B > 4$. Running through values of B we get the following: $B = 5, C = 20$, $B = 6, C = 12$, $B = 8, C = 8$. We stopped at $B = 8$ for the same reason as before.

Now let's look at $A = 5$:
Then we have $\frac{1}{B} + \frac{1}{C} = \frac{3}{10} \Rightarrow \frac{10B}{3B-10} = C$.
C must be an integer and we can see that, for this case, we need $B \geqslant 5$ (due to the equation and the stipulation that $A \leqslant B$). Running through values of B we get the following: $B = 5, C = 10$

Now let's look at $A = 6$:
Oh wait! We did this one already!
Note that if we choose values of A larger than 6, the result will be values of B and C smaller than A, which we cannot have.

We found 10 possible integer set solutions. The largest C can be is 42.
To summarize, our $A - B - C$ solutions were:
3-7-42, 3-8-24, 3-9-18, 3-10-15, 3-12-12, 4-5-20, 4-6-12, 4-8-8, 5-5-10, 6-6-6

4. $\frac{1}{A} + \frac{1}{B} + \frac{1}{C} = \frac{1}{2}$
Multiply both sides by the Least Common Denominator: $2ABC$
$2(BC + AC + AB) = ABC$
Surface Area = Volume

5. Use the results from earlier today! We found all the positive integer solutions to the equation $\frac{1}{A} + \frac{1}{B} + \frac{1}{C} = \frac{1}{2}$. We also know that rectangular box that has integer side lengths (say A, B, C)and it's surface area is equal to its volume satisfied the equation $\frac{1}{A} + \frac{1}{B} + \frac{1}{C} = \frac{1}{2}$
Now put the two together . . .
Possible Boxes:
$3 \times 7 \times 42$, $3 \times 8 \times 24$ $3 \times 9 \times 18$, $3 \times 10 \times 15$, $3 \times 12 \times 12$, $4 \times 5 \times 20$, $4 \times 6 \times 12$, $4 \times 8 \times 8$, $5 \times 5 \times 10$, $6 \times 6 \times 6$

6. This is the set of points (x, y) satisfying $\sqrt{(x-1)^2 + y^2} = \alpha|x+1|$. By experimenting with a dynamic figure or by graphing examples, it seem that this set is

   a. an ellipse if $0 < \alpha < 1$,

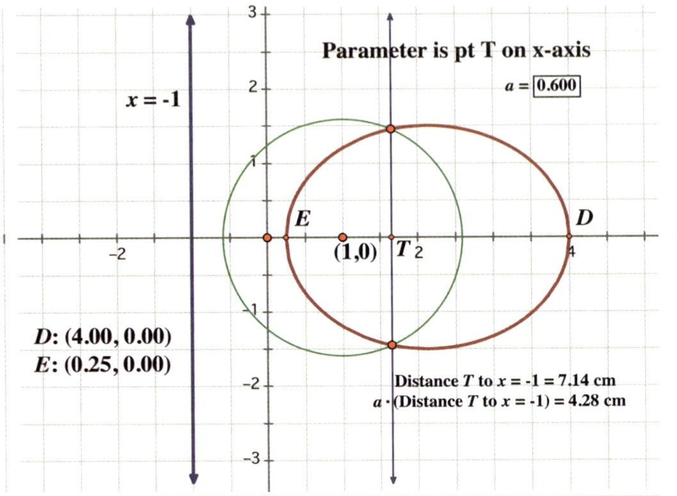

   b. a parabola if $\alpha = 1$,

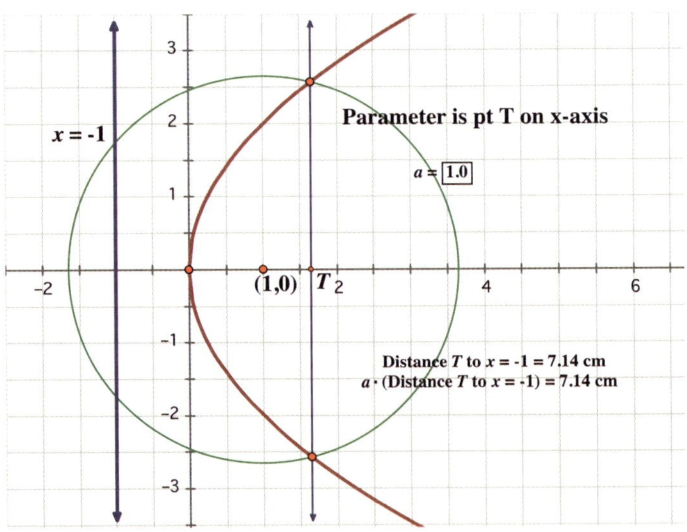

c. and a hyperbola if $1 < \alpha$.

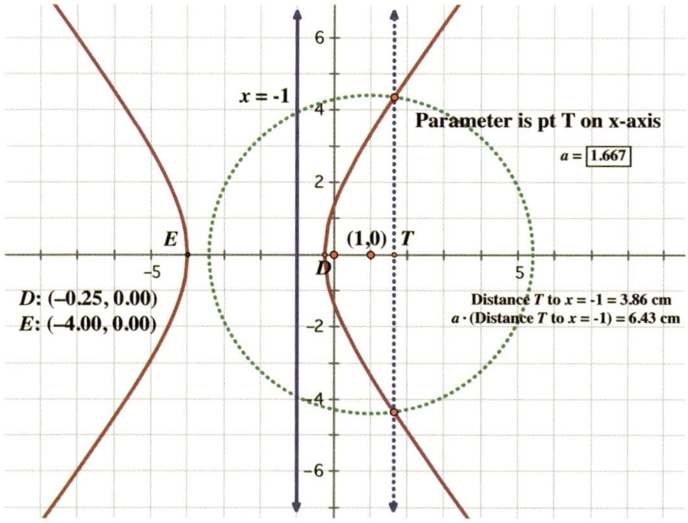

This graphical information can be made more precise with calculations. For example, the set of points on the x-axis, with $y = 0$ satisfies the equation $|x-1| = \alpha|x+1|$. There are two cases:

$$x - 1 = +\alpha(x+1) \Longrightarrow x = \frac{1+\alpha}{1-\alpha}$$

and

$$x - 1 = -\alpha(x+1) \Longrightarrow x = \frac{1-\alpha}{1+\alpha}$$

In the examples shown in the figures, if $\alpha = 3/5$, then $x = 4$ or $x = 1/4$. If $\alpha = 5/3$, then $x = -4$ or $x = -1/4$. If $\alpha = 1$, then one of the equations has no solution and the other has solution $x = 0$.

From the definition it is clear that the figures are symmetric about the x-axis. When $y \neq 0$, then $\sqrt{(x-1)^2} = |x-1| < \alpha|x+1|$.

For points to the "right" of $x = -1$, that is for points with $x + 1 > 0$, this inequality becomes $-\alpha(x+1) < x - 1 < \alpha(x+1)$. This implies both $(1-\alpha)x < 1+\alpha$ and $1-\alpha < (1+\alpha)x$. When $\alpha < 1$, this says that for any (x, y) in the set, then

$$\frac{1-\alpha}{1+\alpha} < x < \frac{1+\alpha}{1-\alpha}$$

This is consistent with the figure of the ellipse. When $\alpha > 1$, the two inequalities only state that $\frac{1-\alpha}{1+\alpha} < x$ when $x + 1 > 0$. Carrying this our further for the case $x + 1 < 0$, we find no points for the case $\alpha < 1$ but the inequality $x < \frac{1+\alpha}{1-\alpha}$ for the case $\alpha > 1$. This is consistent with the figure for the hyperbola.

7. The odd polynomial $z^3 + 4z$ maps the circle either to a path through 0 or to a path that encircles the origin either in one loop or 3 loops, never two loops.

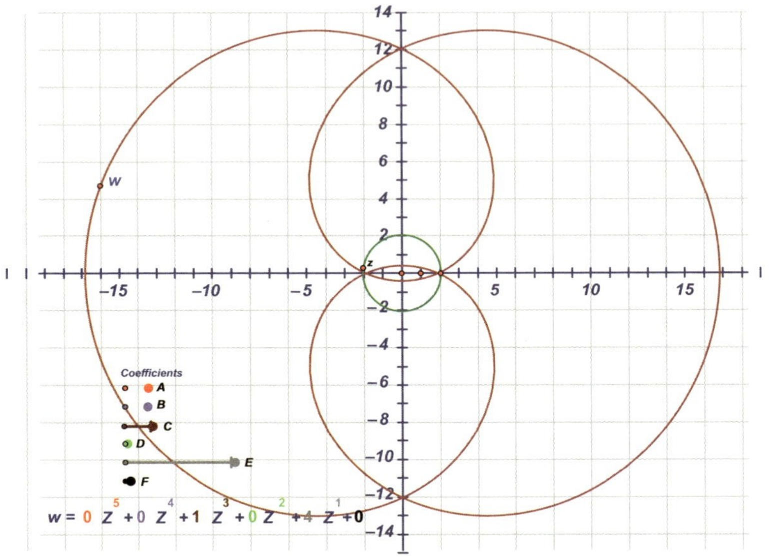

The even polynomial z^4+2z^2 encircles the origin in two loops. As a point z travels one time around the circle, the image point w travels two times around this double loop, since $(-z)^4 + 2(-z)^2 = z^4 + 2z^2$.

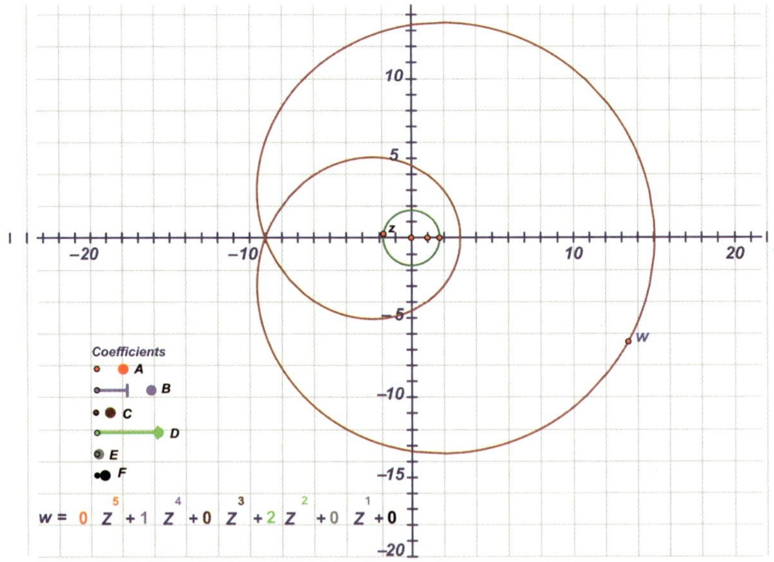

8. To find the solutions of $z^3 + z + 2$ we drag the circle so that the image curve passes through 0, then move the point z on the circle until the image point w lies on 0. One solution is $z = -1$. Another pair of solutions occurs at a double point of the image curve. The values $z = .5 + 1.32i$ and the conjugate value $z = .5 - 1.32i$ are approximate solutions.

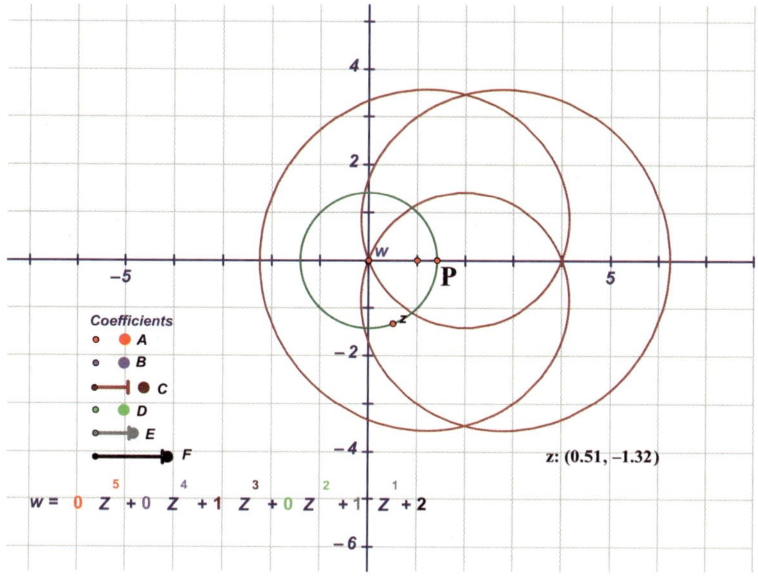

It is possible to find exact solutions using the quadratic formula, since

$$z^3 + z + 2 = (z + 1)(z^2 - z + 2).$$

The exact solution $z = \frac{1+i\sqrt{7}}{2}$ is very close to this approximation.

9. **a.** The complex numbers z with $z^5 = 1$ are the vertices of a regular pentagon. The directions are multiples of $2\pi/5$.

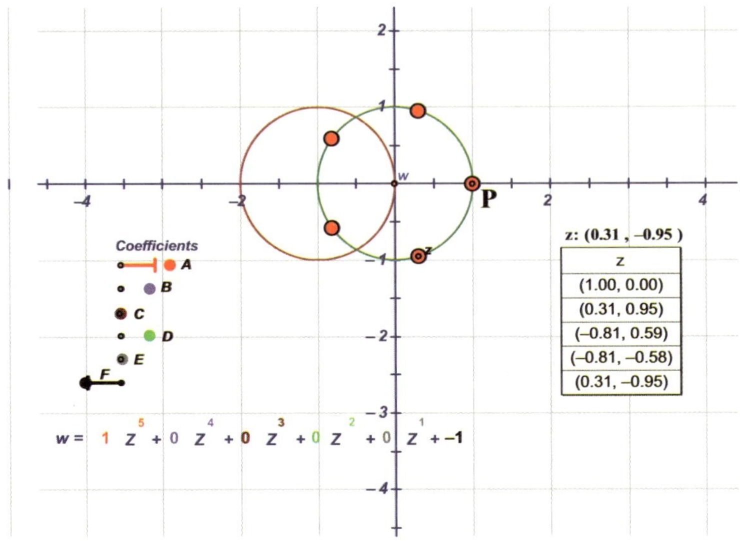

b. The complex numbers z with $z^5 = -1$ are the vertices of a regular pentagon. These are the rotations of the numbers in (a) by 180 degrees about 0, which is the same as multiplying each of the number by -1. To see this, notice that if $w^5 = +1$, then $(-w)^5 = -1$.
These numbers are also square roots of the numbers in (a).

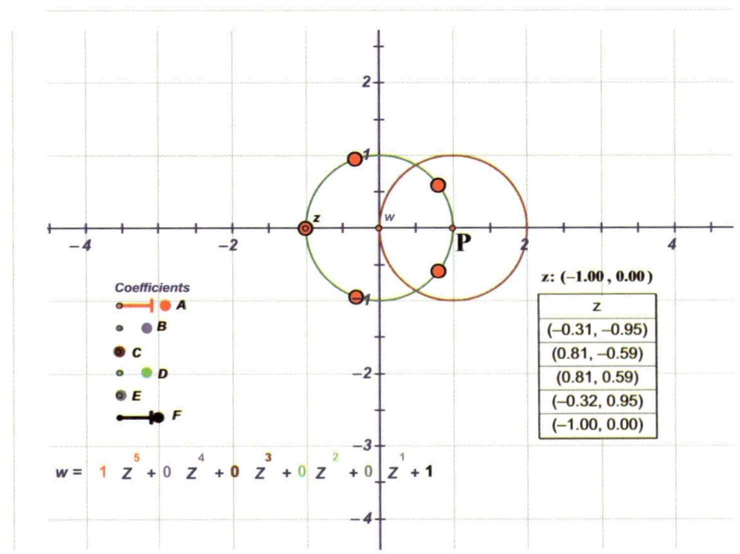

10. Any z from the ten complex numbers in (a) or (b) satisfies $z^{10} = 1$, since $z^{10} = (z^5)^2 = (\pm 1)^2 = 1$. These numbers form a regular decagon.

11. **a.** By really small, we mean that the radius of the circle is small, so that complex numbers z on Mary's circle are very close to the origin in the plane. Thus the points of the output curve of Mary's circle are very close to 5, the constant term of $f(z)$. Therefore the complex number 0 is outside the output curve.

b. When the z circle gets very large the values of the output of $f(z)$ are also large and approximately (percentage-wise) the same as z^2. Therefore the image curve is approximately a large circle wrapping twice around 5. Since the distance is much greater than 5, this curve also wraps twice around 0. When Tina grabs Mary's circle and drags it to become a large circle, the image circle is dragged through 0 twice (or once through a double point) in order for it to become a curve with 0 inside.

12. One thing this expression of "badness" is good for is to conclude that $b = 47$ is the value of b for the line that will minimize badness overall. Combined with the expression in Problem 1, which implies that $b = -4$ for this line, this determines that the equation of the line is $y = -4x + 47$.

But there are other interesting things in this expression. If we specify the value of the y-intercept b, then the line with minimum badness for that given b is the line satisfying the equation $m + \frac{5}{31}b - \frac{111}{31} = 0$. This makes the non-negative first term equal to zero.

Any line satisfying this equation passes through the point $(31/5, 111/5)$. So this point is on the line minimizing badness among all lines. Since we also know that the point $(5, 27)$, obtained by averaging the x and y data, is also on this line, we can deduce the equation of the line of minimum badness as the line through these two points. This again gives the equation of that line: $y = -4x + 47$.

Where does the expression $m + \frac{5}{31}b - \frac{111}{31}$ come from? The formula for the badness is a quadratic polynomial in the variables b and m. If we collect all the terms in m to get $c_1m^2 + (c_2b + c_3)m$ for constants c_i, then by completing the square, this term can be expressed as

$$c_1m^2+(c_2b+c_3)m = c_1(m+\frac{c_2b+c_3}{2c_1})^2-\frac{(c_2b+c_3)^2}{4c_1}$$

This is a complicated expression overall. What numbers are used to compute the coordinates of the point $(31/5, 111/5)$? The number 31 is the average value of the x_i^2 data and the number 111 is the average value of the products x_iy_i. The number 5 is average value of the data x_i. Another way of saying this is that the point is the multiple by 1/25 of the point $(155, 555)$, where 155 is the sum of the squares x_i^2 and 555 is the sum of the products x_iy_i and 25 is the sum of the x_i.

The formula for this second interesting point from this data works for other data as well, as we see in the next problem.

13. The badness function is the sum of the squares $(y_i - mx_i - b)^2$ for each of the data points (x_i, y_i). Since

$$\begin{aligned}&(y_i - mx_i - b)^2\\&\quad = y_i^2 + m^2x_i^2 + 1b^2 - 2mx_iy_i - 2by_i + 2mbx_i\end{aligned}$$

the coefficients of the variables are the sums

$$\begin{aligned}1+1+1+4 &= 4\\x_1+x_2+x_3+x_4 &= 16\\y_1+y_2+y_3+y_4 &= 20\\x_1^2+x_2^2+x_3^2+x_4^2 &= 84\\x_1y_1+x_2y_2+x_3y_3+x_4y_4 &= 96\\y_1^2+y_2^2+y_3^2+y_4^2 &= 138\end{aligned}$$

so the badness formula is

$$138 + 84m^2 + 4b^2 - 192m - 40b + 32mb$$

From the earlier experiments we expect that if we fix any slope m and find the line of least badness, then the line will pass through the balance point, which is the average or mean of the four data points. This point is $\frac{1}{4}(16, 20) = (4, 5)$. The equation for all the lines through this point is $4m + b - 5 = 0$, so following the earlier example, this suggests that the badness can be written as a sum $4(b + 4m - 5)^2 + q(m)$, where $q(m)$ is a quadratic expression in m (the 4 is chosen because it is the coefficient of b^2 in the badness formula).

We can compute $q(x)$ as the difference of the badness and $4(b + 4m - 5)^2$ or by substituting $b = 5 - 4m$ for b into the badness formula. This gives

$$\begin{aligned} q(b) &= 20m^2 - 32m + 38 \\ &= 20(m - \frac{16}{20})^2 + 38 - \frac{256}{20} \\ &= 20(m - \frac{4}{5})^2 + \frac{126}{5} \end{aligned}$$

and the badness formula is

$$4(b + 4m - 5)^2 + 20(m - \frac{4}{5})^2 + \frac{126}{5}$$

For a second formula modeled on the formula in Problem 12, with the sum of a squared expression in b and another square of a linear expression, we see that the point through which all the lines pass has x-coordinate equal to the average of the x_i^2 data divided by the average of the x_i. The the y-coordinate is the average of the $x_i y_i$ divided by the average of the x_i. For this data, this point is $(21/4, 24/4) = (21/4, 6)$.

Thus $\frac{64}{21}(b + \frac{21}{4} - 6)^2 + p(b)$, with $p(b)$ quadratic in b, should be another formula for the badness.

Computing as before, we find this second formula for badness:

$$\frac{64}{21}(b + \frac{21}{4}m - 6)^2 + \frac{20}{21}(b - \frac{9}{5})^2 + \frac{126}{5}$$

It should be noted that in developing these formulas, we have found two points on the line that minimizes the badness overall, namely $(4, 5)$ and $(21/4, 6)$. The line through these two points has the equation $y = \frac{4}{5}x + \frac{9}{5}$. This is the line that minimizes badness. It is the regression line for the data.

14. Experiment suggests that the segment PX bisects the angle APB.

15. If X is any point on segment AB, then taking AX and BX as the bases of the triangles AXP and BXP, the height of each triangle is the same h – the distance of P from line AB. Thus the ratio of the areas of AXP and BXP is the ratio

$$\frac{\text{Area } AXP}{\text{Area } BXP} = \frac{|AX|h/2}{|BX|h/2} = \frac{AX}{BX}$$

There is another way to compute areas of the two triangles AXP and BXP. Take the base of each triangle to be the common side PX. Then the heights of the triangles are the lengths of perpendicular segments from A and from B to the line PX. Denoting the feet of these perpendiculars by A' and B' as in the figure, the triangles PAA' and PBB' are right triangles.

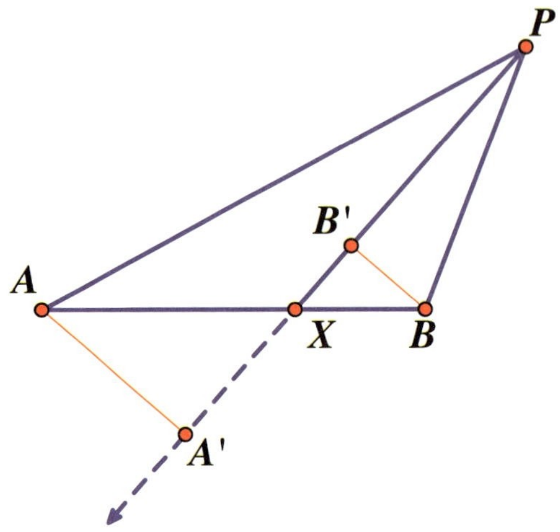

If PX is the angle bisector of angle APB, then the angle APX is congruent to angle BPX and the two right triangles are similar. This similarity implies that $AA'/BB' = AP/BP$. So

$$\frac{\text{Area } AXP}{\text{Area } BXP} = \frac{|PX|\,|AA'|/2}{|PX|\,|BB'|/2} = \frac{AP}{BP}$$

Putting this together with the previous equation for the ratio of the areas, we conclude that if X is the intersection of the angle bisector of angle APB with segment AB, then $\frac{AX}{BX} = \frac{AP}{BP}$.

The converse of this is true as well. If we begin with a point X on segment AB and a point P such that $\frac{AX}{BX} = \frac{AP}{BP}$, then the intersection of the angle bisector of angle APB with segment AB is a point X' on segment AB so that $\frac{AX'}{BX'} = \frac{AP}{BP} = \frac{AX}{BX}$. But there is only one point on the segment with this ratio, so $X' = X$.

16. For any $k > 0$, there are two points T on line AB for which $AT/AT = k$. To see this, consider a correspondence of the line AB with the real number line, with A, B, and T corresponding to real numbers a, b, and t. The equation of ratios of distances $|\frac{t-a}{b-t}| = k$ is equivalent to these two equations:

$$\frac{t-a}{b-t} = k \quad \text{and} \quad \frac{t-a}{b-t} = -k$$

The solutions are

$$x = \frac{a + kb}{1 + k} \quad \text{and} \quad y = \frac{a - kb}{1 - k}$$

with the exception that for $k = 1$ there is no solution y and x is the midpoint of the interval $[a, b]$. Let X be the point corresponding to number x and Y the point corresponding to y. Then X is on the interval AB and Y is outside the interval.

Experiment suggests that the segment PY is perpendicular to the line PX. It is also true that PY bisects the *exterior* angle APB.

17. The demonstration that line PY is perpendicular to line PX will refer to this figure.

The point X is the point on segment AB so that $AX/XB = PA/PB$; in Problem 15 it was shown that PX is the angle bisector of angle APB. The point Z is defined to be the intersection of line AB with the line through P perpendicular to PX. (This point exists except in the special case where X is the midpoint of AB.

The point E is any point on line AP so that P is between E and A. Also, we assume that the triangle is labeled so that larger of the two segments PA and PB is PA so that B is between A and Z.

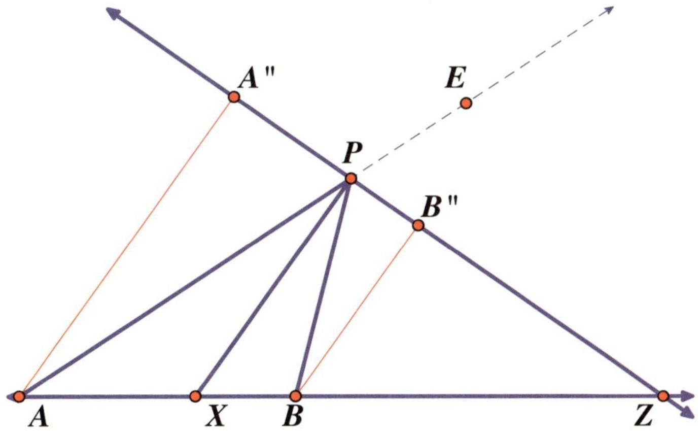

Drop segments AA'' and BB'' perpendicular to line PZ as in the figure.

Since PX is the angle bisector, $\measuredangle XPA = \measuredangle XPB = \theta$.

The angle $\measuredangle APA'' = 90\,^\circ - \theta$ is the complement of XPA since XPA'' is a right angle. Likewise, $\measuredangle BPB'' = 90\,^\circ - \theta$. Thus $\measuredangle APA'' = \measuredangle BPB''$.

This implies that the right triangles PAA'' and PBB'' are similar. Therefore $AA''/BB'' = PA/PB$.

Now we can show that $\frac{AZ}{BZ} = \frac{AP}{BP}$ by – as before – computing the ratio of the areas of the triangles AZP and BZP in two different ways. First, we take the base of each triangle on line AB; both triangles have the same height h.

$$\frac{\text{Area } AZP}{\text{Area } BZP} = \frac{|AZ|h/2}{|BZ|h/2} = \frac{AZ}{BZ}$$

Second, we take the base of each triangle to be the common side PZ. The heights of these triangles are the lengths of the perpendicular segments AA'' and BB''. So

$$\frac{\text{Area } AZP}{\text{Area } BZP} = \frac{|PZ|\,|AA''|/2}{|PZ|\,|BB''|/2} = \frac{AA''}{BB''} = \frac{PA}{PB}$$

Since $AZ/BZ = PA/PB$ and Z is exterior to segment AB, the point Z must b the point Y introduced in Problem 16. This shows that PY is perpendicular to PX.

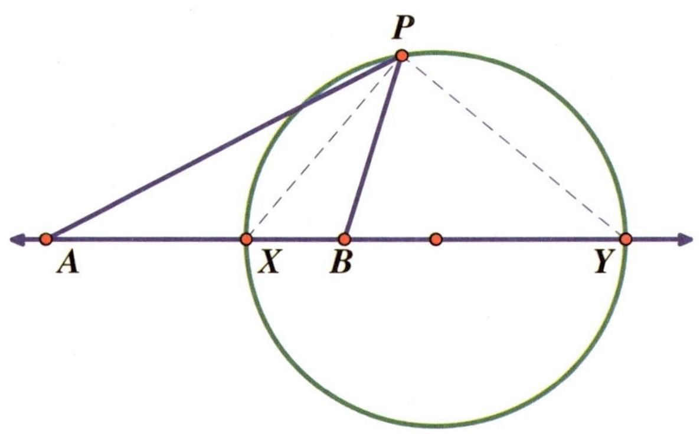

A key point here is that the points X and Y only depend on the ratio PA/PB, not on any other property of P. Thus, given two points A and B and a real $k > 0$, the set of points Q so that $AQ/QB = k$ intersects the line AB at points X and Y (as above) and for any other Q in the set, the lines QX and QY are perpendicular.

This means that the locus of Q is contained in the circle with diameter XY (or the perpendicular bisector of AB when $k = 1$). Since these circles for distinct values of k are disjoint, and since every point Q must belong to the circle for $k = QA/QB$, the locus of Q for a particular k is equal to the circle with diameter XY for the value of $k = QA/QB$.

Problem Set 13

Opener

We can see that when fitting regular polygons together around a vertex, that the sum of the interior angles of the polygons around the vertex must sum to 360. The order of the polygons can matter if you are wanting to create different patterns.
Let the number n represent a regular n-gon (polygon with n sides). Also note that the interior angles (in degrees) of a regular n-gon can be found using $\frac{180(n-2)}{n}$.
So the angles of a 4-gon (or square) are $\frac{180(4-2)}{4} = 90°$ (as if we're surprised!)
The angles of a 5-gon (or pentagon) are $\frac{180(5-2)}{5} = 108°$
And so on . . .
The possible ways to arrange three n-gons around a vertex are:
6-6-6 (*Three 6-gons*, as given)
3-7-42 (*One 3-gon, one 7-gon, and one 42-gon*)
3-8-24 (*One 3-gon, one 8-gon, and one 24-gon*)
3-9-18 (*One 3-gon, one 9-gon, and one 18-gon*)
3-10-15 (*One 3-gon, one 10-gon, and one 15-gon*)
3-12-12 (*One 3-gon and two 12-gons*)
4-5-20 (*One 4-gon, one 5-gon, and one 20-gon*)
4-6-12 (*One 4-gon, one 6-gon, and one 12-gon*)
4-8-8 (*One 4-gon and two 8-gons*)
5-5-10 (*Two 5-gons and a 10-gon*)
Wait a minute . . . weren't these the same solutions to the problem in the previous problem set, which tied to the solutions to $\frac{1}{A} + \frac{1}{B} + \frac{1}{C} = \frac{1}{2}$, which also tied to the rectangular boxes?? Hmmm . . . why? How?
Let's say we have an A-gon, a B-gon, and a C-gon that can all fit around a vertex with no gaps.
So the angles would add up
$\frac{180(A-2)}{A} + \frac{180(B-2)}{B} + \frac{180(C-2)}{C} = 360$
Giving $\frac{A-2}{A} + \frac{B-2}{B} + \frac{C-2}{C} = 2$
$1 - \frac{2}{A} + 1 - \frac{2}{B} + 1 - \frac{2}{C} = 2$
$\frac{2}{A} + \frac{2}{B} + \frac{2}{C} = -1$
$\frac{1}{A} + \frac{1}{B} + \frac{1}{C} = \frac{1}{2}$
WHAT!

1. The possible ways to arrange four n-gons around a vertex are:
3-3-4-12 (*Two 3-gons, a 4-gon, and a 12-gon*)
3-3-6-6 (*Two 3-gons and two 6-gons*)
3-4-4-6 (*One 3-gon, two 4-gons, and a 6-gon*)
4-4-4-4 (*Four 4-gons*)
Again, note that there can be variations with the above arrangements to create different patterns.

2. Use the ideas from Problem Set 12 to solve this problem. In Problem Set 12 we solved $\frac{1}{A} + \frac{1}{B} + \frac{1}{C} = \frac{1}{2}$
This time, we are solving $\frac{1}{G} + \frac{1}{A} + \frac{1}{B} + \frac{1}{E} = 1$.
Suppose G is the smallest value: what could it be? G must either be 3 or 4: we require $G \geqslant 3$, and if $G > 4$ there will not be enough to add to 1.
What if $G = 3$? The equation then becomes
$\frac{1}{A} + \frac{1}{B} + \frac{1}{E} = \frac{2}{3}$
Solving using the methods from Problem Set 12, we get:
Possible values for B and E are
$A = 3, B = 4, E = 12$, $A = 3, B = 6, E = 6$, $A = 4, B = 4, E = 6$
If $G = 4$ the only solution is $A = 4, B = 4, E = 4$.
So, all the solutions for $G - A - B - E$ are:
3-3-6-6, 3-3-4-12, 3-4-4-6, 4-4-4-4
Check these solutions out: 3-3-6-6, 3-3-4-12, 3-4-4-6, 4-4-4-4. Do you see any relation to the 4 n-gons around a vertex problem? Sweet!

3. **5 n-gons:**
We want to fit 5 n-gons around a vertex, so our options are going to start being limited.
$\frac{360°}{5} = 72$ and there is not an n-gon with such interior angle measure, so we cannot do the same shape all the way around. Let's start with triangles and see what happens. If we fit 4 triangles around, we have $360° - 240° = 120°$ remaining, so we can fit in a hexagon. So 3-3-3-3-6 is one possibility.
If we fit 3 triangles around, we have $360° - 180° = 180°$ remaining for two more n-gons. Two 4-gons will fit perfectly. So 3-3-3-4-4 is the other possibility.
If we use only two triangles, we have 300° remaining to divide among four n-gons, and nothing will work.

6 n-gons:

The only option for 6 n-gons is to use all triangles 3-3-3-3-3-3.

7 n-gons:

Cannot fit 7 n-gons around a vertex. The smallest n-gon is a 3-gon, and seven 3-gons together have an angle sum greater than 360°.

Alternately, for five regular polygons you could solve the equation $\frac{1}{A} + \frac{1}{B} + \frac{1}{C} + \frac{1}{D} + \frac{1}{E} = \frac{3}{2}$ with the restriction that A through E are integers 3 or larger. The same solutions should be available. The equation $\frac{1}{A} + \frac{1}{B} + \frac{1}{C} + \frac{1}{D} + \frac{1}{E} + \frac{1}{F} = 2$ is used for six regular polygons.

4. The best solution adds two additional points in the interior of the square, parallel to two sides, creating 120-degree angles to the vertices. These additional points, sometimes called *Steiner points*, allow the total path length to be minimized.

Other solutions, such as connecting three vertices, or connecting the diagonals, are not as efficient as the solution with Steiner points.

5. A right circular cylinder with radius r and height h has surface area $A = 2\pi r^2 + 2\pi rh = 2\pi r(r+h)$ and volume $V = \pi r^2 h$. If $A = V$, then $2\pi r(r+h) = \pi r^2 h$. This implies $2(r+h) = rh$, which can be rearranged as

$$\frac{1}{r} + \frac{1}{h} = \frac{1}{2}$$

This is an equation that we have seen before. Two solutions are $r = 3$, $h = 6$ and $h = 3$, $r = 6$. Another solution is $r = 4$, $h = 4$.

6. The area and volume are not enough to determine the radius and height of a right circular cylinder.

For example, the cylinder with radius $r_1 = 2$ and height $h_1 = 12$ has volume equal to $\pi {r_1}^2 h_1 = 48\pi$ and area $2\pi r_1(r_1 + h_1) = 56\pi$. But the cylinder with radius $r_2 = 4$ and height $h_2 = 3$ also has volume equal $\pi {r_2}^2 h_2 = 48\pi$ and area $2\pi r_2(r_2 + h_2) = 56\pi$.

A second pair of examples is the cylinder with radius $r_1 = 1$ and height $h_1 = 6$ and the cylinder with radius $r_2 = 2$ and height $h_2 = 3/2$. For each one the volume is 6π and the surface area is 14π.

Here is one way to generate such examples. For a right circular cylinder, let $\tilde{A} = \text{Area}/\pi = 2r(r+h)$ and let $\tilde{V} = \text{Volume}/\pi = r^2 h$. Suppose that $\tilde{A}$ and

$\tilde{V}$ are given values. How can we find r and h so that the cylinder has these values of $\tilde{A}$ and $\tilde{V}$?

We can solve the volume formula for h and then substitute into the area formula. Solving $h = \tilde{V}/r^2$, then

$$\tilde{A} = 2r(r + \frac{\tilde{V}}{r^2}) = \frac{2r(r^3 + \tilde{V})}{r^2} = \frac{2(r^3 + \tilde{V})}{r}$$

This is a cubic equation in r:

$$r^3 - \frac{\tilde{A}}{2}r + \tilde{V} = 0$$

If r is any positive solution of this polynomial equation, then one can solve for $h = \tilde{V}/r^2$ and obtain the dimensions of a cylinder with the given values of $\tilde{A}$ and $\tilde{V}$.

But to find two distinct positive integer roots $r_1 = a$ and $r_2 = b$ of this equation, it is possible to start with a and b and work backwards. Let the third root of the polynomial be c. Since the coefficient of r^2 in the polynomial $(r - a)(r - b)(r - c)$ is 0, this means that $a + b + c = 0$, or $c = -(a + b)$.

Also, the constant term $\tilde{V} = abc = -ab(a + b)$. Finally, the coefficient of r is $ab + ac + bc = ab - (a + b)^2 = -\tilde{A}/2$.

Thus we can pick any two radii $r_1 = a$ and $r_2 = b$. Then, from the polynomial $(r - a)(r - b)(r + (a + b)) = r^3 - \tilde{A}r/2 + \tilde{V}$, compute values of $\tilde{A}$ and $\tilde{V}$ that will be equal for two cylinders with these radii and heights $h_1 = \tilde{V}/a^2$ and $h_2 = \tilde{V}/b^2$. To find the area and volume, multiply $\tilde{A}$ and $\tilde{V}$ by π.

7. The image of the "magnitude circle" circle passes through 0 at a double point of the curve. The point z on the circle can be moved to two positions that move the image w to 0. One approximate value, measured from the coordinate measure in the software, is $z = 1+1.73i$ and the other is $z = 1.01 - 1.73i$. Allowing for measurement error and considering that the roots should be conjugate, this seems pretty close. Of course if the point were not to observe the dynamic nature of the image curve, one could find the exact roots with the quadratic formula.

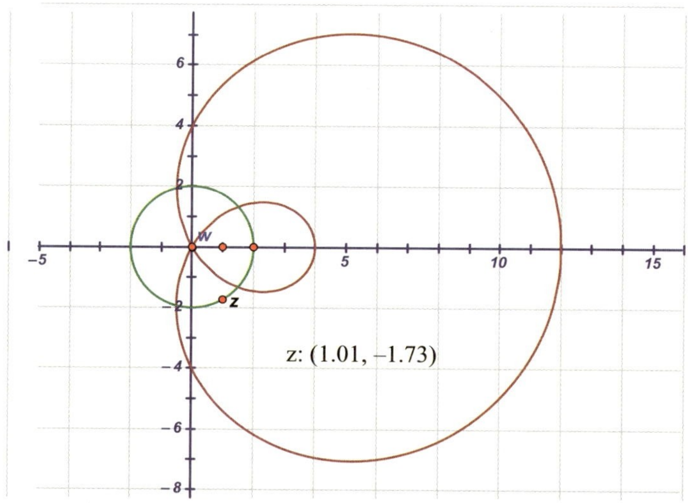

8. Instantly estimate $f(1/100)$ as 3 and $f(100)$ as 10^6, or one million.

9. a. When $|z|$ is very small, for example on the circle $|z| = 1/100$, then $f(z)$ is very close to 3.

   b. When $|z|$ is very large, for example on the circle $|z| = 100$, then the output of $f(z)$ is close to the circle $|z| = 10^6 = 1,000,000$. It is never farther away than $6|z| + 3 = 603$. Also, as z goes around the circle one time, the image $w = f(z)$ goes around the big circle three times as z^3 does. In other words, as the angle of z moves from 0 to 2π the angle of w moves – not quite at a constant rate – from 0 to 6π.

   c. As the radius r of the z circle $|z| = r$ increases, the image curve, which is a triple loop, must pass through 0 to reach the curve described in (b). This is visually very convincing in Sketchpad. For another way of putting this, if one looks at the net increase of the direction angle of w, it begins with a net increase of 0 when the image curve is almost constant near $w = 3$ for small z. But this net increase of angle is a continuously varying integer multiple of 2π as z varies (hence constant!), except that a discontinuity occurs when the image curve passes through 0. Thus the curve must pass through 0 at least once, and unless it passes through a multiple point, it will pass through 3 times to get to 6π.

10. **a.** $f(1/100) \approxeq -4$
b. $f(-1/100) \approxeq -4$
c. $f(100) \approxeq 10^6$
d. $f(-100) \approxeq -10^6$

11. The answers and reasons are the same, except that for small z the image curve is very close to -4.

12. For any complex polynomial $f(z)$ of degree $n \geqslant 1$, there is at least one z for which $f(z) = 0$.

A convincing reason for this is the same as the one given in Problem 9. If the constant term of $f(z)$ is a_0, then for very small r, the image curve of the circle $|z| = r$ is very close to a_0. Unless $a_0 = 0$ (in which case $z = 0$ is a root and we are done!), this curve is contained in a disk arround a_0 some distance from 0.

But if r is very large, the image of image curve of the circle $|z| = r$ is very close to the path of z^n as it travels n times around the circle $|z| = r^n$. For the same reasoning as in Problem 9, since the image curve varies continuously from small path not containing zero to an almost circle path that travels n times around 0, it must pass through 0 at least once. In addition to the visual image, again one may reason about the net increase in the direction angle of w as it goes around the image curve. It begins with 0 net increase for small z and ends with $2n\pi$ for large z.

13. The exterior angle of a regular polling with n sides is $360/n$ measured in degrees or $2\pi/n$ measured in radians. We will use degrees, but the algebra is the same in either measure.

If the exterior angle is $360/n$, then the interior angle is the difference with a straight angle, which is

$$180 - \frac{360}{n} = 360(\frac{1}{2} - \frac{1}{n})$$

If we have four regular polygons fitting together at a point, then the sum of the interior angles must be 360. If the number of side of the four polygons is m, n, p, q, then summing, we get

$$360(\frac{1}{2} - \frac{1}{m} + \frac{1}{2} - \frac{1}{n} + \frac{1}{2} - \frac{1}{p} + \frac{1}{2} - \frac{1}{q})$$
$$= 360(2 - \frac{1}{m} - \frac{1}{n} - \frac{1}{p} - \frac{1}{q}) = 360$$

Dividing by 360 we get

$$\frac{1}{m} + \frac{1}{n} + \frac{1}{p} + \frac{1}{q} = 1$$

The positive integer solutions for this equation were found in Problem 2.

14. **a.**

$$\frac{1}{6} + \frac{1}{6} + \frac{1}{4} = \frac{7}{12} = \frac{1}{2} + \frac{1}{12}$$

So the polygons fall $\frac{1}{12}$ short of fitting together.

b. This excess of $\frac{1}{12}$ corresponds to 30 "missing" degrees. there would be $720 = 2 \times 360$ total missing degrees for 24 vertices.

c. There is a polyhedron with two regular hexagons and one square at each vertex. This might be thought of as a tiling of the sphere. The polyhedron can be constructed by staring with a regular octahedron, which is made of 8 triangles meeting 4 at each vertex.

If one takes an equilateral triangle of side s and cuts off a corner equilateral triangle of side $s/3$ from each of the 3 corners, the remaining figure is a regular hexagon. If one does this cutting to each of the triangles of the octahedron, the resulting object consists of 8 hexagons and six square-shaped holes at each of the vertices of the octahedron, which we can fill in with 6 squares. The six squares have 24 vertices total, so from (b) we see that the total of the "missing" degrees = 720.

But wait! For the original octahedron, each vertex is formed from 4 triangles, so the "missing" degrees equal 120. And with 6 vertices, $6 \times 120 = 720$ again! Hmm.

Problem Set 14

Opener

If the magnitude of z is small, we get something like:

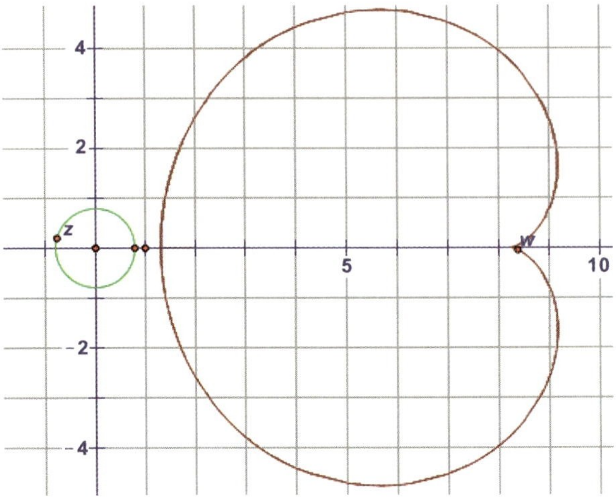

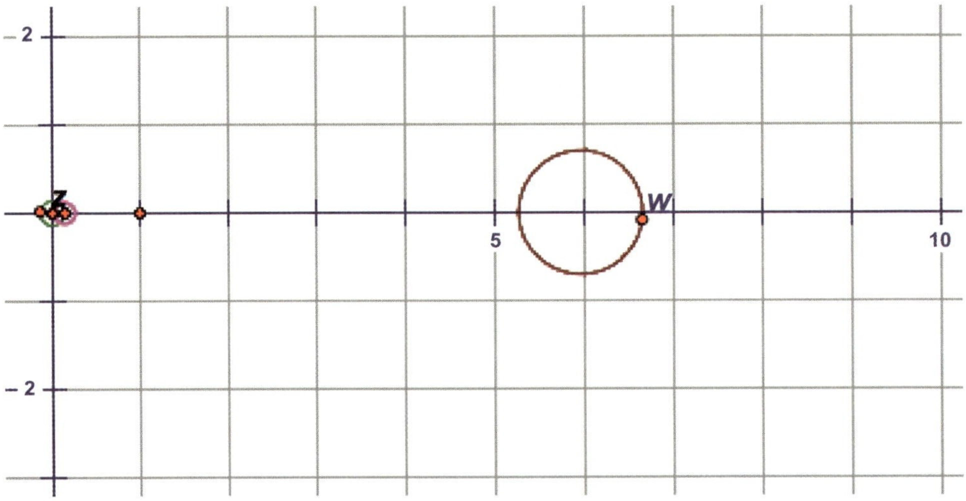

What is f(z) doing as z gets tiny. Hmmm.... But what if z is big?

Check it out:

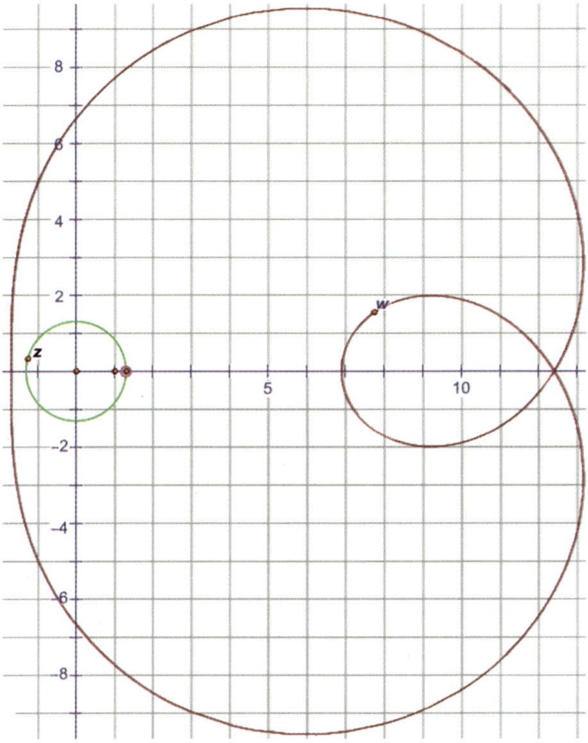
8
6
4
2
z
w
5
10
-2
-4
-6
-8

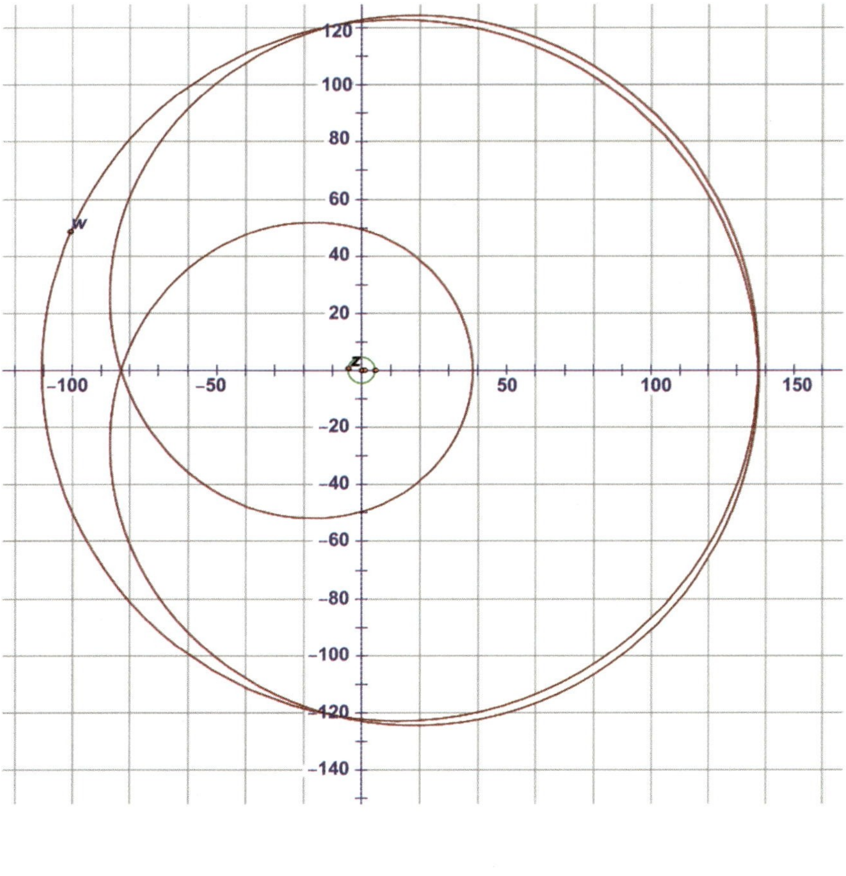
120
100
80
60
w
40
20
z
-100
-50
50
100
150
-20
-40
-60
-80
-100
-120
-140

It doesn't take much to make $f(z)$ to explode!

We can see that the graph of $f(z)$ passes through the origin and thus intersects a circle centered about the origin. So there is a magnitude of z such that $f(z) = 0$. Check it out. If $z = 1, -2, 3$, we definitely get $f(z) = 0$.

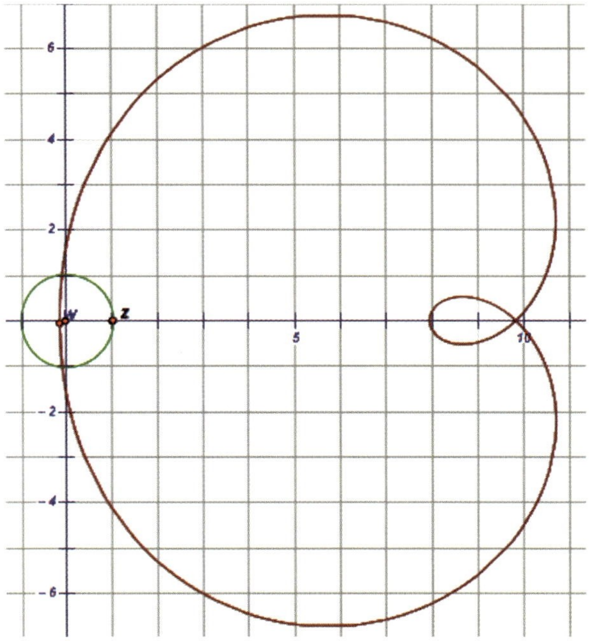

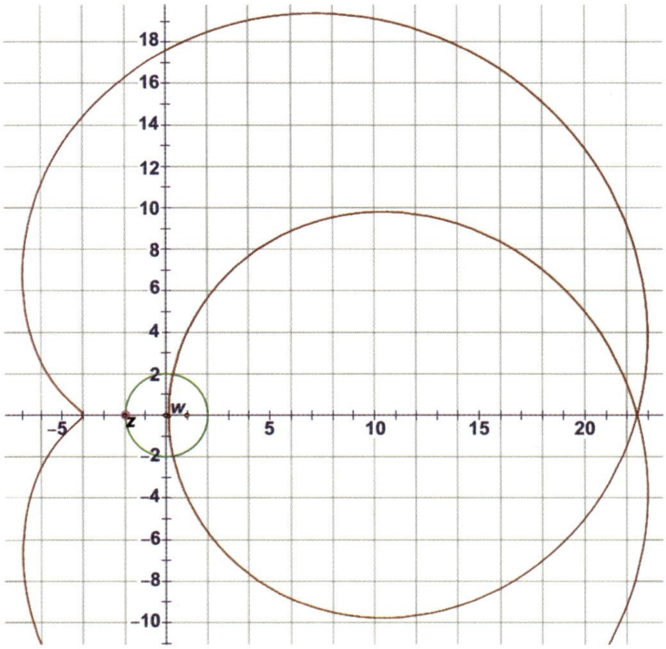

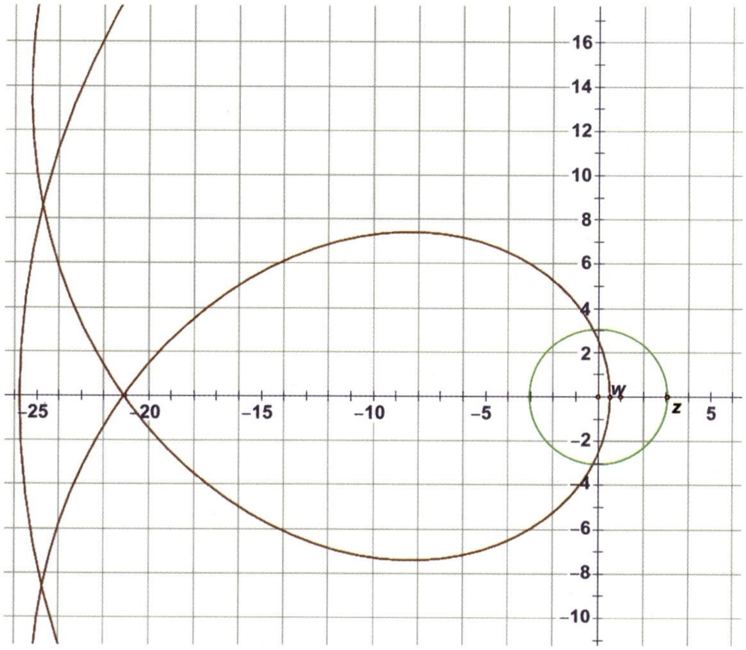

1. Use `http://tinyurl.com/comppolygsp`. Play around with the sliders. What do you notice? Will things change if the polynomial is of higher degree?

2. Consider the picture below. The angles meeting at vertex X are all right angles.

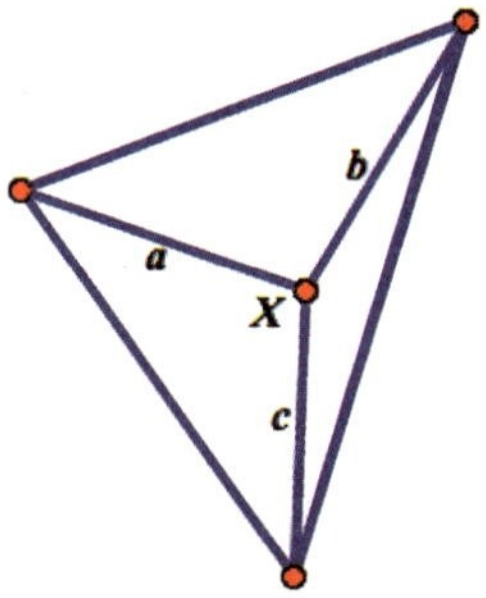

We can see that the area for each of these 3 faces meeting at vertex X are $\frac{ab}{2}, \frac{ac}{2}, \frac{bc}{2}$.
The sides that form the bottom face are the hypotenuses from the three right triangles on the top of our tetrahedron. Thus the sides are $\sqrt{a^2+b^2}, \sqrt{a^2+c^2}, \sqrt{b^2+c^2}$
Go back to Problem Set 9 or use Heron's formula if you must to find the area.
Have fun multiplying! You should get:
$A = \frac{1}{2}\sqrt{\left(\frac{ab}{2}\right)^2 + \left(\frac{ac}{2}\right)^2 + \left(\frac{bc}{2}\right)^2}$
Note that this shape is a *Trirectangular Tetrahedron*.

3. See if the picture below sparks any ideas for this proof. Note that the surrounding triangles are all equilateral triangles.

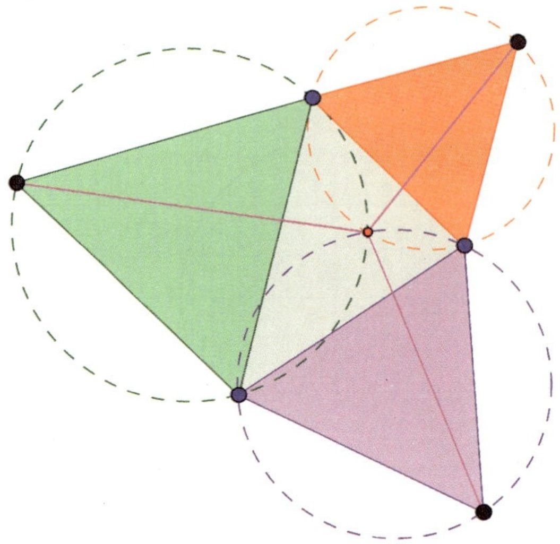

4. I've got all kinds of triangles!

   Triangle Set #1

   Triangle #1: side lengths 20, 21, 29; Triangle #2: side lengths 17, 25, 28

   Both have area $A = 210$ and perimeter $P = 70$.

   Triangle Set #2

   Triangle #1: side lengths 25, 34, 39; Triangle #2: side lengths 24, 37, 37; Triangle #3: side lengths 29, 29, 40

   Both have area $A = 420$ and perimeter $P = 98$.

5. (a) Check it out:

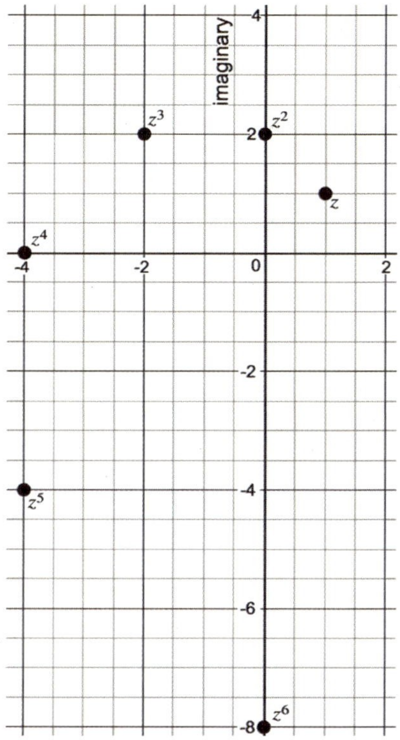

(b) And some more spiral action! Yessss!

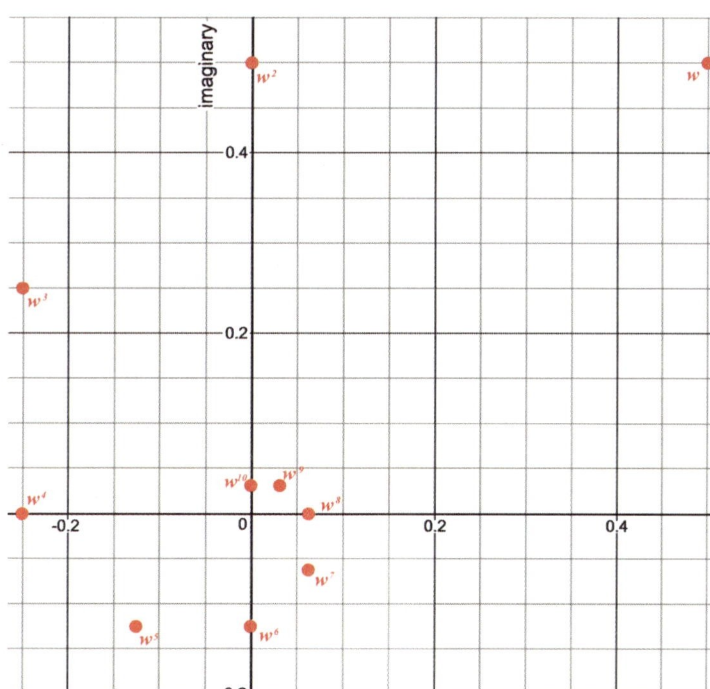

(c) The infinite sum is approaching $1 + i$. Whoa!! And we have more spiral action!

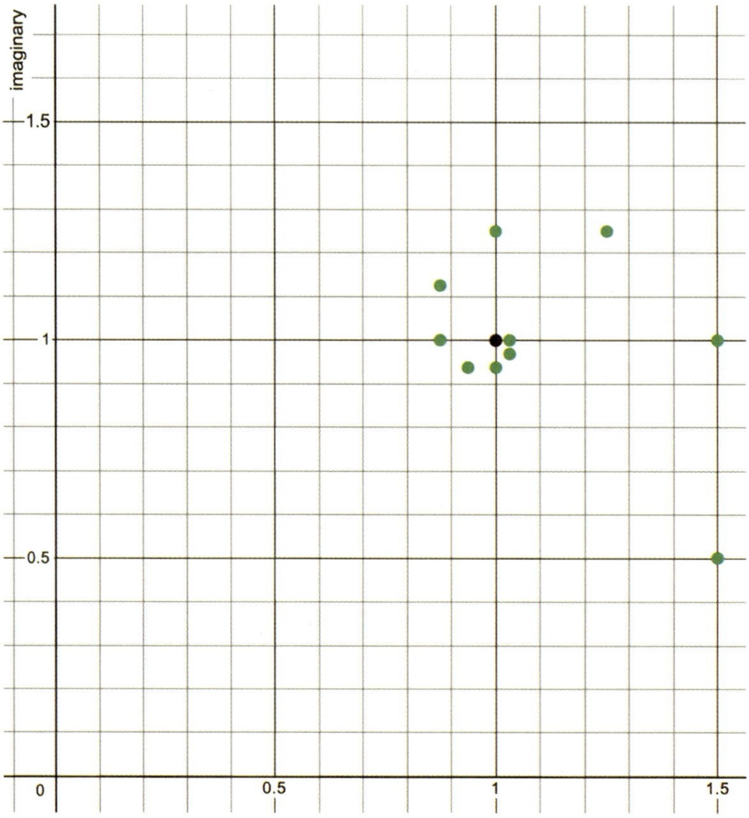

6. We know that if $n = 1$ we get $kx_1 = 2$

   If $n = 2$ we get $kx_1x_2 = 2(x_1 + x_2)$

   If $n = 3$ we get $kx_1x_2x_3 = 2(x_2x_3 + x_1x_3 + x_1x_2)$

   This can tie into the Volume and Area box problems. Neato!

7. Three lines must be used to connect everybody. Two lines will not be sufficient, as either a pair of people will be disconnected from the other pair or one person will be disconnected from the other three. Now the question is: Which three lines?

   We should definitely use the line between P and L and the line between A and S, because they are the shortest segments. The diagonals are definitely the longest segments, so it's better to choose the segment joining P and A or the segment joining L and S.

General quadrilateral?! Sheesh... You've got this.

5 points? Of course...

8. Go back to the opener from Problem Set 13. Which of those can tessellate the plane?

   Hexagons. Definitely hexagons. What else? Does the 4-gon + 6-gon + 12-gon work? Hmmm....

9. (a) 3-gon: Let $s = 4\sqrt{3}$. Then $A = 12\sqrt{3} = P$.
   (b) 4-gon: Let $s = 4$. Then $A = 16 = P$.
   (c) 6-gon: Let $s = 4\frac{\sqrt{3}}{3}$. Then $A = 8\sqrt{3} = P$.
   (d) 8-gon: Let $s = 4\sqrt{2} - 4$. Then $A = 32\sqrt{2} - 32 = P$.
   (e) 12-gon: Let $s = 4(2 - \sqrt{3}$. Then $A = 48(2 - \sqrt{3}) = P$.
   (f) n-gon: What generalizations have you been able to make so far? Can you break the shapes up into triangles? How does that help you?

10. Well, we have to solve $2ab + 2ac + 2bc = A$ (where A is surface area) and $abc = V$ (where V is volume). If a, b, c are integers, what does that mean for A and V?

    If we know A and V, we can also set up their ratio equation for solving: $\frac{A}{V} = \frac{2}{c} + \frac{2}{b} + \frac{2}{a}$

11. No! Give it a try with the triangle with side lengths 5, 30, 29 (Area is $A = 72$).

12. Let's do this in general. Let $m = \frac{a}{b}$. Then the line through $(0, -1)$ with slope $m = \frac{a}{b}$ is $y = \frac{a}{b}x - 1$.

    This line with intersect the hyperbola at a second point for some values of m (notice it will not intersect if $m = 1$). Funky!

    If the line is intersecting the hyperbola, then the x and y values coincide at that point.

    So we can write:
    $(\frac{a}{b}x - 1)^2 - x^2 = 1$, which leads us to: $\left(\frac{a^2}{b^2} - 1\right)x^2 - \frac{2a}{b}x = 0$
    $\Rightarrow x = 0, \frac{2ab}{a^2 - b^2}$.
    Then we get $y = -1, \frac{a^2+b^2}{a^2-b^2}$
    $(0, -1)$ and $\left(\frac{2ab}{a^2-b^2}, \frac{a^2+b^2}{a^2-b^2}\right)$.
    Now play around with different slopes!

Notice what happens if $a = b$. Hmmm.... Such points lie on that funky line with slope $m = 1$. Interesting. What if $m = -1$. Cool.

13. $\sqrt{i} = \frac{\sqrt{2}}{2} + i\frac{\sqrt{2}}{2}$
$\sqrt[3]{i} = -i$
But wait... $\left(\frac{-\sqrt{3}}{2} + \frac{i}{2}\right)^3 = i$. But wait! $\left(\frac{\sqrt{3}}{2} + \frac{i}{2}\right)^3$ is also i. So I am guessing that there is another one for $\sqrt{i}$. Could it be... $-\frac{\sqrt{2}}{2} - i\frac{\sqrt{2}}{2}$. Yes indeed.
Based on this pattern, I bet there are 6 to look for for $\sqrt[6]{i}$.
Notice that $\left(\frac{\sqrt{2}}{2} - i\frac{\sqrt{2}}{2}\right)^6 = i$ and $\left(\frac{-\sqrt{2}}{2} + i\frac{\sqrt{2}}{2}\right)^6 = i$. Can you find more?

14. (a) $\left(\frac{1}{2} - \frac{\sqrt{3}}{2}, \frac{1}{2} + \frac{\sqrt{3}}{2}\right)$
(b) $\frac{3}{2} + \sqrt{3}, -1 + \frac{3\sqrt{3}}{2}$
(c) $\frac{5}{2} + \frac{5\sqrt{3}}{2}i$
(d) $\frac{1}{2}(a - b\sqrt{3}) + \frac{i}{2}(a\sqrt{3} + b)$

15. (a) You should be able to find two points that work. One of them is $\left(3.5 + \frac{\sqrt{3}}{2}, 4.5 - \frac{\sqrt{3}}{2}\right)$ Use what you know about distance between two points and equilateral triangles. Think on this. If you are still stuck, check out some of the ideas from part (d).
(b) You should be able to find two points that work. One of them is $\left(3.5 + \sqrt{3}, \frac{3\sqrt{3}}{2}\right)$
(c) This one isn't so bad! It's cool because both points lie on a horizontal line. Use your knowledge of equilateral triangles, distance, etc. $\left(\frac{6x+5}{2}, 2 + \frac{5\sqrt{3}}{2}\right)$
Check out the picture of the triangles for (a), (b), and (c). I let $x = 2$ for part (c).

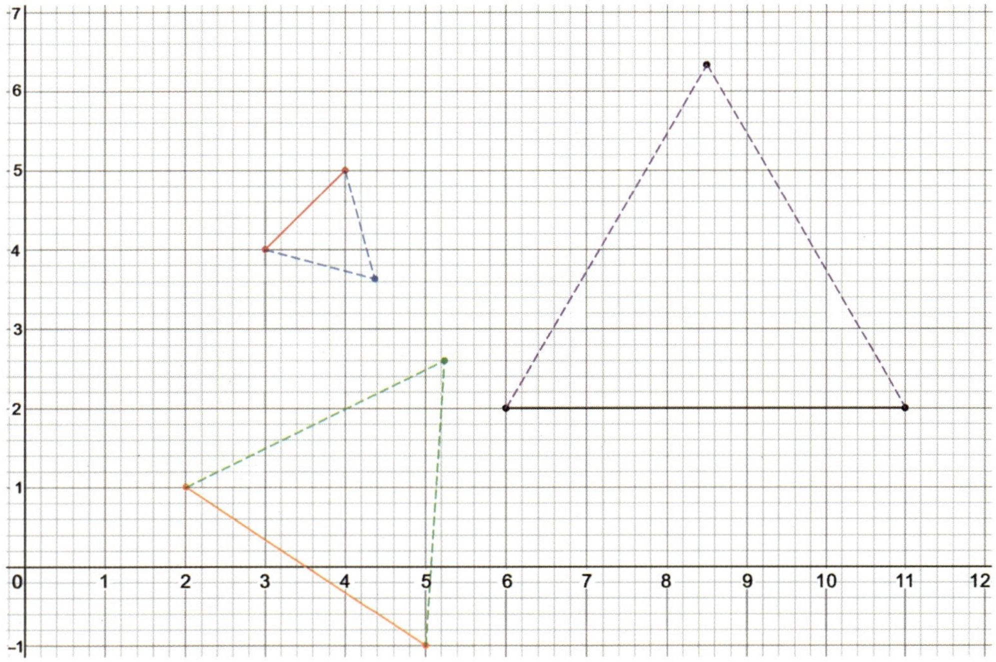

(d) What now?

Use what you know about the distance formula and equilateral triangles. Suppose the 3rd point that needs to be found is (A, B) and the distance between each point be s.

Known:

$(x - A)^2 + (y - B)^2 = x^2$

$(w - A)^2 + (z - B)^2 = x^2$

$\left(\frac{x+w}{2} - A\right) + \left(\frac{y+z}{2} - B\right) = \frac{3}{4}s^2$

Fun with algebra will give:

$$B = \frac{y+z}{2} - \frac{\sqrt{3}s}{2\sqrt{\left(\frac{y-z}{x-w}\right)^2 + 1}}$$

and

$$A = \frac{x+w}{2} - \frac{(y-z)(y+z-2B)}{2(x-w)}$$